Building
as an
Economic Process

Building
as an
Economic Process

An Introduction
to
Building Economics

Ranko Bon

Massachusetts Institute of Technology

Prentice Hall, Englewood Cliffs, New Jersey 07632

Library of Congress Cataloging-in-Publication Data

BON, RANKO
   Building as an economic process.

   Bibliography: p.
   Includes index.
      1. Building—Economic aspects.   2. Real estate
   management.   3. Capital.   I. Title.   II. Title:
   Building economics.
   HD1382.5.B66   1989   690'.068'1   88-23187
   ISBN 0-13-086166-9

Editorial/production supervision
   and interior design: *Jean Lapidus*
Manufacturing Buyer: *Mary Ann Gloriande*

 © 1989 by Prentice-Hall, Inc.
A Division of Simon & Schuster
Englewood Cliffs, New Jersey 07632

Printed in the United States of America

10 9 8 7 6 5 4 3 2 1

ISBN 0-13-086166-9

PRENTICE-HALL INTERNATIONAL (UK) LIMITED, *London*
PRENTICE-HALL OF AUSTRALIA PTY. LIMITED, *Sydney*
PRENTICE-HALL CANADA INC., *Toronto*
PRENTICE-HALL HISPANOAMERICANA, S.A., *Mexico*
PRENTICE-HALL OF INDIA PRIVATE LIMITED, *New Delhi*
PRENTICE-HALL OF JAPAN, INC., *Tokyo*
SIMON & SCHUSTER ASIA PTE. LTD., *Singapore*
EDITORA PRENTICE-HALL DO BRASIL, LTDA., *Rio de Janeiro*

For my parents

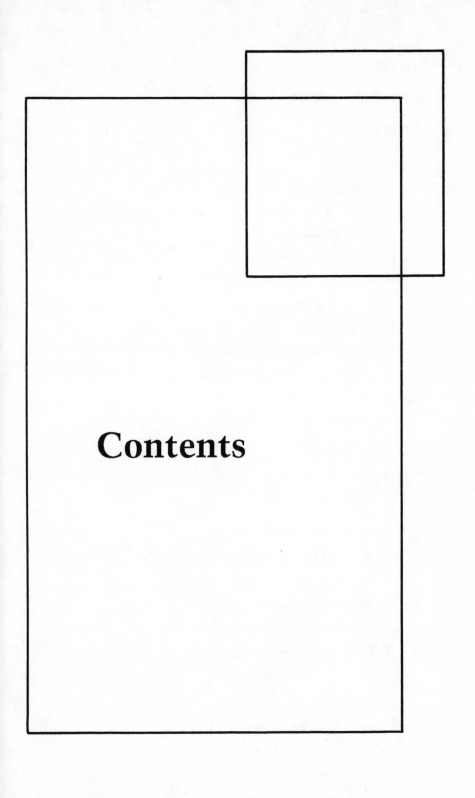

Contents

3 BUILDING PROCESS   55

4 BUSINESS AND BUILDING CYCLES   80

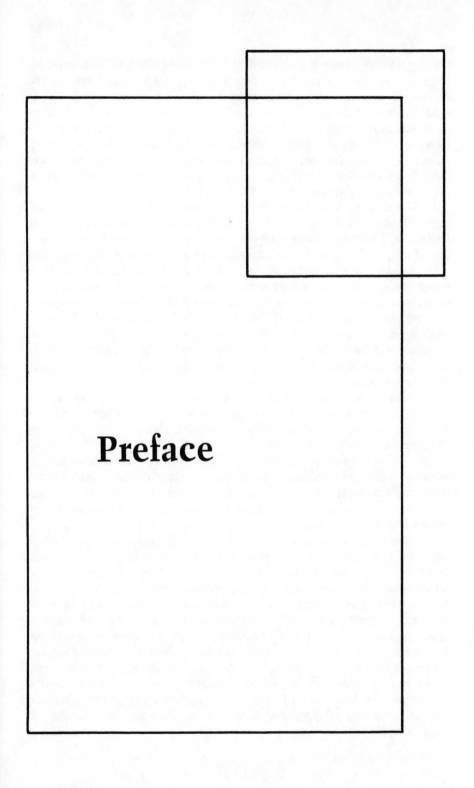

Preface

Building economics emerged as a distinct field in mid-1970's, induced by the so-called energy crisis. More than a decade later, it is still in its infancy. The economics profession does not recognize it as a field in its own right. Yet, in most economies, buildings represent a large part of the capital stock, as well as the annual capital expenditures. Why then is building economics developing at such a sluggish pace, and what are the reasons for its lack of professional recognition? In my opinion, both questions have a simple answer: the field lacks theoretical foundations. My objective is to offer a first step toward a theoretical framework for building economics.

This small book went through many a transformation since its conception in the summer of 1984. From the very beginning of my search for a unifying framework, I have been convinced that an economic understanding of time is central in building economics. Building takes time, and building life cycles span many decades. As time and change are intrinsically bound, building activity is fraught with genuine or unavoidable uncertainty. The economic agents engaged in the building process therefore learn as they act in real time. Their expectations are continually revised as they gain new knowledge. My search for a conceptual framework sensitive to the centrality of time in economic affairs ultimately led me to the so-called Austrian school. The Austrian theories of capital, money, and economic fluctuations, interrelated as they are, offered a rich environment for subsequent discoveries.

Many ideas offered here have been discussed in my classes in building economics in the Department of Architecture at MIT. My students, most of whom are working toward degrees in architecture, civil engineering, and city planning, have often felt uncomfortable with standard presentations of microeconomic and macroeconomic principles, that is, with neoclassical orthodoxy. It fails to address many issues characteristic of the building realm. My increasingly frequent excursions into Austrian themes resonated well with my students, and that encouraged me to pursue these ideas even further and to give this book its present form.

I hasten to add that my leanings toward the Austrian school, which originated in Vienna in the late nineteenth century, are by no means doctrinaire in nature. I am not concerned with what Menger—the progenitor of the school—really said, nor with assorted squabbles whether or not Hicks or Shackle—two great economists whose work is close to this school—are true Austrians. Simply stated, I believe that the Austrian school has contributed to economic thought much that I find useful for the development of building economics. Of course, this is not to say that other schools have nothing to offer. Far from it. As will be increasingly apparent in the chapters that follow, I have drawn heavily from other schools of economics, including the neoclassical mainstream. For these reasons I do not feel compelled to engage in a systematic comparison of various approaches to specific problems discussed, let alone in disputes on doctrinal matters.

In this book I explore some key issues of building economics in the context of the theory of capital and the theory of economic fluctuations or business cycles. Again, my main purpose is to provide the foundations of a theoretical framework that will inform further development of building economics. Thus far building economics has applied standard investment decision criteria to buildings as a special class of capital assets. However, this approach is largely *ad hoc,* and the theoretical foundation is still lacking. Here I offer a first step toward a consistent framework for an explanation of economizing behavior in the building arena. Without a clear link to economic theory, building economics will neither develop beyond a narrow domain of project evaluation, including capital budgeting and benefit-cost analysis, nor gain recognition as a field of economics proper.

The main idea behind this book is rather simple. Heretofore, building economics has focused on forecasting the economic consequences of a building decision on the basis of ever more extensive historical data about individual buildings and their components. However, I believe that another key task of building economics is to assist the decision makers concerned with building economy in their day-to-day operations involving the entire real property holdings at their disposal. The problem with the old focus of the field is twofold. On the one hand, economic history of building activity is an important area of study in its own right, but the practical significance of building economics hinges on the field's sensitivity to the anticipation of continual economic change facing decision makers. On the other hand, economic forecasting is not convincing over the extended periods of time associated with the building life cycle. The focus should therefore shift from investment decisions to decisions concerning the use of capital. Such an active orientation to problems of building economy of course requires theoretical underpinnings.

The shift of emphasis from building investment to building utilization and operation in part reflects a long-term shift in the construction sector itself. First, in most industrialized countries the share of construction in gross national product is decreasing. Second, the share of new construction in total construction is decreasing as well, while the share of so-called maintenance and repair construction—including rebuilding in all its forms—is increasing. A significant proportion of the real property holdings will remain in the "portfolios" of building owners well beyond their present business horizons. Real property management, as distinct from development, is gaining momentum in professional circles.

There is also a conceptual reason for this change of perspective. Schools of economic thought that have focused on capital utilization rather than on the level of investment, such as the Austrian school, are being rediscovered by researchers seeking conceptual frameworks corresponding more closely to the actual experience of building professionals.

In this light, a few words about the title of the book are in order. The word "building" has two meanings, because it is both a common noun and a verbal noun. It refers both to the building process and the final outcome of that process. However, the life of the "completed" building itself is an ongoing process, continually adapting through building activity to the changing economic process that gave it birth. The expression "building process" has two meanings, too. It refers both to the process of building and to the resulting building understood as a process. The two meanings of the expression are thus doubly intertwined. In the last analysis, the "ambiguity" inherent in the notion of building process exposes more than it hides, as I will endeavor to show in the pages that follow.

This book is intended for building economists, as well as other building professionals concerned with building economy—real estate developers and managers, architects, quantity surveyors and cost estimators, civil engineers, and facilities managers. I hope that building clients will find it useful, as well. In fact, building clients are increasingly concerned with coordination of the building process; they must play an active role in it to ensure that their needs will be met. Building economics may help unify the building field, presently characterized by fragmentation and avoidable conflict. If this book succeeds in bringing us a step closer to the integration of the building field, I will consider it to have been well worth the effort.

All the technical aspects of the presentation are placed in rather lengthy notes following each chapter. These notes are pointers to the literature that I recommend for further reading. Several diagrams are provided to illustrate the points made in the text. There are no mathematical arguments anyplace in the book. The only exception is an appendix, where some formalisms for symbol manipulation are briefly presented. The book is thus accessible to a wide readership spanning the building professions. However, some background in economics, or at least a familiarity with—and an appreciation of—the way economists think, is necessary to fully comprehend the arguments offered here.

I am painfully aware of the potential danger involved in an attempt to speak to so many diverse professions at the same time. What may appear to be the best balance between economics and the building professions, on the one hand, and between scholarly and professional concerns, on the other, may in fact miss the real reader. To speak to everyone is perhaps to speak to no one. Nevertheless, this is the risk I am willing to take. At this stage of its development, building economics requires the bridging of many separate domains. That task cannot be avoided any longer.

In these chapters, I will first examine the economic underpinnings of building activity. Second, I will introduce some essential concepts in the theory of capital. This is the theoretical backbone of the book. Third, I will elaborate the concept of building process, and its components—the processes of planning, design, construction, operation, and utilization of buildings. I will analyze the entire building process in the context of changing economic conditions. Fourth, I will focus on building and business cycles. The emphasis will be placed on economic changes that affect the economy as a whole. Fifth, I will consider the research tasks of building economics and its potential courses of development. This is the key to the entire book. In a sense, the first four chapters provide the theoretical justification for the last chapter. There, I will offer a very personal view of what is to be done, tempered by experience in sponsored research and consulting for several private and public organizations of considerable size. It is meant to be neither exhaustive nor authoritative, for the development of building economics requires a collective effort.

A caveat about the structure of the book may be useful: the reader should keep in mind that the five chapters forming this book are very much interrelated. A concept introduced in one chapter is often developed—and sometimes modified—in a subsequent chapter. For this reason, the reader should be prepared to suspend his or her disbelief until the last chapter. Considering the book's size, this is not an unreasonable request.

Cambridge, Massachusetts   *Ranko Bon*
May 1988

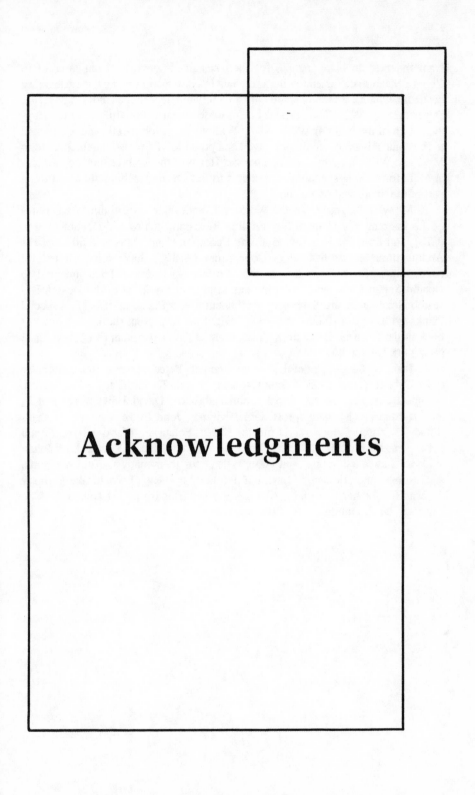

Acknowledgments

I am indebted to many people for their sustained support. I am grateful to George Macomber for endowing the chair in construction management bearing his name, which I presently share with a faculty member in the Civil Engineering Department at MIT. The purpose of this endowment is to help reduce the fragmentation of the building professions. Without the scholarly allowance attached to the chair this book would not have been possible. I am also indebted to John Myer and William Porter, past and present Heads of the Architecture Department at MIT, for making special arrangements to lighten my teaching duties during a part of the time spent on this book.

Many colleagues from the Working Commission on Building Economics of the International Council for Building Research Studies and Documentation (CIB), and from the Panel on Building Economics and Industry Studies of the Architectural Research Centers Consortium (ARCC), have provided helpful comments since the inception of this project. In addition, I am indebted to Donald Schön for several opportunities to present my ideas in the Design Research Seminar at the School of Architecture and Planning at MIT. Special thanks go to Mario Rizzo for having invited me to present the ideas from this book in the Austrian Economics Colloquium at the Department of Economics at New York University.

I am especially grateful to John Bennett, Peter Brandon, Jan Bröchner, Peter Cebon, Young Chai, Joseph Csillaghy, Guido Dandri, Patrice Derrington, Sivaguru Ganesan, Steven Groàk, John Habraken, David Hawk, Tom Heath, Robert Johnson, Michael Joroff, Israel Kirzner, John Lowe, Harold Marshall, Sidney Newton, Roberto Pietroforte, Karen Polenske, Mario Rizzo, Gyula Sebestyen, Klara Szöke, Herman Tempelmans Plat, Francis Ventre, Lars Johan Walden, Lawrence White, and Alan Wilson for their many useful comments, suggestions, and criticisms. Last, but definitely not least, I would like to thank Teresa Hill for her editorial assistance and her friendship. Of course, I alone claim all the remaining faults of this book.

It is by reason of the existence of durable equipment that the economic future is linked to the present.

> John Maynard Keynes, *The General Theory of Employment, Interest, and Money* (1936)

The future is not there to be discovered, but must be created.

> George Shackle, *Decision Order and Time in Human Affairs* (1969)

It all begins in the mind's eye.

> Courtland Collier and Don Halperin, *Construction Financing* (1984)

1

Building Activity

INTRODUCTION

The fragmentation of the building professions, the building process, and the built environment is one of the fundamental problems of the building industry today. The underlying interconnections within the building field require systematic elucidation. Building economics provides a unifying framework for the study of building as a purposeful and rational human activity.

This is not to say that all manner of whims and fancies manifested in the built environment are excluded from consideration here. Even the most extravagant and ostensibly irrational projects must be conducted so that the available means correspond to the ends, no matter how these ends were chosen. Economics is a science of means, not of ends. Rationality consists in the consistent pursuit of one's own purposes.[1]

The emerging field of building economics is gradually penetrating both professional practice and academic curricula in architectural and engineering design. Although many designers still tend to be content with ready-to-use economic formulas embedded in proliferating software packages, a growing minority is certainly aware that formulas can be of help only to those who understand their meanings. In this spirit, we will explore a simple question: What can building economics contribute to building design, construction, and management? As this book demonstrates, this contribution centers on the notion of time. Indeed, the formulas concerning building economy are most sensitive to the assumptions about time-related phenomena.

Why is a dynamic conception of building activity important? Buildings undergo continual alterations as they are adapted to the needs of their owners. In turn, these needs evolve in response to continually changing economic conditions. As these changes cannot be fully foreseen, buildings must be designed and constructed so they may be adapted to a wide range of conditions that may be encountered in the underlying economic process. Adaptability must not be confused with physical plasticity, however. As Lynch [1972: 112] argues, "[u]seful adaptability is not eternal neutral plasticity but rather the current maintenance of a continuing capability to respond to change so as to achieve changing objectives." Adaptability requires continual attention on the part of building management. Therefore, building manageability should be a key design criterion from the vantage point of a building owner.

Building economics may not remedy all that ails the building industry, but it may be used to diagnose what is happening around us. An economic understanding of building activity as a whole is a precondition for further improvements in economizing the use of building resources. One of the main tasks of building economics is to explain the economic causes and consequences of human action, as manifested in the built environment.

We will introduce in this chapter some fundamental concepts and propositions to be used throughout this book for an economic explanation of building activity. These concepts will be further developed and substantiated in the subsequent chapters. First, we will discuss the connection between building economics and the theory of capital, concentrating on buildings as production goods. Second, we will introduce the notion of building process, fundamental to building economics. Third, the main economic agents involved in the building process will be presented. Fourth, the notion of human action will be applied to building activity. Fifth, building activity will be presented in the context of production plans, in which buildings will be addressed as parts of capital combinations underlying production processes. Sixth, we will explore the connection between buildings and land, two components of real property.

BUILDING ECONOMICS AND THE THEORY OF CAPITAL

Building economics is about economizing the use of scarce resources throughout the life cycle of a building from conception to demolition. This includes human resources needed for building management. The management of building services—consisting primarily of the use of space and the amenities provided in it—represents one of the scarce resources that needs to be economized. A building should be understood as part of a capital combination, possibly involving other buildings, dedicated to a specific purpose. This purpose, embedded in a production plan, may change over time. This is a dynamic economizing problem, which may be defined as that of allocating scarce resources among competing ends over an interval of time. The problem of capital arises as soon as we allow time to pass, as capital links the economic future to the present [Keynes, 1964: 146].

To illustrate the importance of time for the understanding of buildings as capital goods, it is sufficient to note that the delivery of building services cannot be compressed in time at our will. As Georgescu-Roegen [1971: 225] argues, "some things can be consumed at once but others are durable because their consumption requires duration." Contrasting candy and hotel rooms, Georgescu-Roegen [1971: 226] writes:

> If the count shows that a box contains twenty candies, we can make twenty youngsters happy now or tomorrow, or some today and others tomorrow, and so on. But if an engineer tells us that one hotel room will probably last one thousand days more, we cannot make one thousand roomless tourists happy *now*. We can only make one happy today, a second tomorrow, and so on, until the room collapses.

Therefore, the use of building services requires duration. As we will argue in Chap. 2, this difference characterizes not only capital versus noncapital goods, but also fixed versus circulating or working capital, respectively.

Because buildings are a significant part of long-lived capital goods, we may consider building economics as a special branch of the theory of capital, concerning the pattern of investment in and use of capital goods under changing economic conditions.[2] Propositions from the theory of capital may be adapted to address problems in the use of building resources.

The theory of capital has been one of the most difficult and controversial areas of economic theory.[3] This book is based on the Austrian school of economics. Members of this school have been actively involved in numerous polemics about the nature of capital.[4] However, the objective here is to lay the groundwork for building economics, not to join the controversy between schools of thought. It will be shown in Chap. 2 that the Austrian propositions concerning capital goods in general are useful to the study of building economics. The reasons will become increasingly apparent as we proceed. In this endeavor, the emphasis will be placed on buildings as *sui generis* production goods, which undergo continual change as they are being used in the production process.

Some definitions of capital goods include residential buildings. Here, residential buildings will be treated as durable consumption goods. Similarly, income-generating property of all kinds, including apartment and office buildings, will be treated as durable items of wealth. This subject will be taken up in the next chapter.

BUILDING PROCESS

Building process is a notion central to building economics, as we will argue in greater detail in Chap. 3. Buildings are creatures of time, and thus of change. People participating in the building process continually modify their knowledge, as well as their preferences and expectations.

Clearly, the longer a process takes, the more likely it is that unexpected change will occur. As Lachmann [1978a: xiv] argues, "[t]he theory of capital is a dynamic theory, not merely because many capital goods are durable, but because the changes in use which these durable capital goods undergo during their lifetime reflect the acquisition and transmission of knowledge." The more durable a capital good, the more pertinent is this argument. The role of expectations is especially important in this context. In Lachmann's [1986: 17–18] words:

> In general, the more durable the goods traded in a market, the more important are expectations for it. In capital goods markets and those for durable consumer goods they matter more than in those for simple consumer goods. In a modern market economy, in which there are markets for permanent assets, such as shares in capital combinations which outlast the individual capital goods forming part of them, expectations matter more in these markets than in those for single capital goods, first- or second-hand. Here expectations matter not only because such capital combinations are regarded as sources of income streams extending into an infinite future, but also for another reason often not fully appreciated: [. . .] all permanent

assets that have ever been created still are, in principle, at least potentially "in the market."

This is why time, a crucial element in all economic processes, is especially important to building economics. It is the ultimate scarce resource for all human endeavors.[5] Even more important, economic agents learn throughout the building process, continually modifying their preferences and expectations as they interact. The time dimension is therefore at the root of building activity, as the building life cycle may span entire historical epochs.

When a building process begins, the people involved realize they cannot know what will happen in the future. They know that the future is unknowable. However, they must act, proceeding from moment to moment though their knowledge can never be complete. As they act, these economic agents create a future they can never completely control.

In this book, we will focus on buildings in use. An economic understanding of building utilization and operation is essential for an explanation of the building process as a whole. Buildings are designed and erected to provide useful services to their owners, whose needs change as economic conditions change. Seen from this perspective, the designer's and builder's job is never done. This insight is easily forgotten in the hectic process of planning, design, and construction, when a building is likely to be perceived as a final and immutable product.

ECONOMIC AGENTS

The sequence of actions leading to the ultimate form of a building deserves special attention. Starting from a production plan, the client specifies what building services are needed and determines the budget. In a sense, the client desires building services, rather than a building itself. This distinction is important because these services may be provided in several alternative ways, some of which may not require building in the sense of "bricks and mortar." However, in addition to the building budget, the designer usually receives from the client the building program only. Such a program specifies the client's needs in terms of areas required for different functions, together with the functional relationships between these areas. The designer transforms the client's specifications into images of physical form and instructions for the builder. Under typical contractual arrangements, the builder is engaged after the design has been completed. Consequently, the designer often operates without specific information on the building technology available to the builder. This tends to favor conservative building practices [Cassimatis, 1969: 118–19]. Interpreting the designer's directions, the builder assembles the appropriate materials, equipment, and personnel. Each economic agent is moved by a unique set of needs and incentives, but underlying the entire building process are the purposes of the client—often the ultimate consumer of building services.[6]

If the client is unsatisfied, or if new conditions alter the client's needs, these communications may be repeated several times before actual construction begins. As the process unfolds, it becomes increasingly difficult to change the building each agent has in mind. At some point the building becomes reality. The client then operates and manages it as part of a capital combination informed by a particular production plan. If the plan changes, building may begin anew. During its life cycle, a building may undergo this process several times. We are concerned with the economic underpinnings and consequences, both intended and unintended, of this process.

In market economies the building process may be thought of as a *sui generis* market process. Each construction project represents a temporary market with three principal economic agents: the client, the designer, and the builder.[7] Of course, many less visible economic agents are also involved, as will be shown later, but the interplay between these three economic agents is central. Their relationships reflect the specific costs, prices, and contractual arrangements established for this particular temporary market and span the construction phase of the building's life cycle. In a sense, the language of the market is the only language all the participants in the building process share. As Turin [1966: 13] observes, "[p]erhaps the only common factor underlying the whole [building] process is the economic one." Because each economic agent has distinct objectives and methods, the building process requires continual coordination and negotiation. The market process is the framework within which each economic agent economizes the use of his or her resources.

Building economics overlaps construction economics because the building process includes the construction process itself. At present, construction economics is perhaps the most developed area of building economics.[8] However, building economics integrates the construction phase into a more comprehensive conceptual framework. Building economics is about why—rather than how—a building is constructed in a particular way. The construction process and the underlying building technology need to be understood in the context of building utilization and operation—the primary foci of building economics.

HUMAN ACTION

Buildings are conceived in human minds.[9] They are intrinsically bound to human purposes. To understand the building process, it is essential to understand how an economic agent thinks about it and its ultimate result. People act to shape an unknown and unknowable future in accordance with their purposes. The notion of action already implies the uncertainty of the future. Choosing and acting would be impossible in a perfectly predictable world. Natural science makes it possible to foretell the consequences of definite actions under definite conditions, but it does not render the future predictable. Human acts of choice are inherently unpredictable [Mises, 1966: 105].

Although unknowable, the future is not unimaginable [Lachmann, 1978b: 3]. Because buildings may render useful services for decades or even centuries, the importance of human imagination and creativity in the building process is heightened. This applies especially to the building owner and designer, as well as to their relationship. Building economics is not about buildings per se, as objects of economic activity, but about the choices and values brought to the process by people, the subjects of economic activity.[10]

Building is an intersubjective or social process. However, the social group involved in the building process is not an acting being. We should bear in mind that only individuals act [Mises, 1978: 78]. Collaboration between individuals is an aspect of human action. Unlike most consumption goods, produced for an impersonal market, durable production goods tend to be designed and built in direct collaboration between many economic agents. This is especially true of buildings used for production purposes, which are by and large custom-designed and built. How individuals interact during the building process changes from one historical epoch to another. According to Goldthwaite [1980: 124]:

> There is a direct relation between consumer and supplier that separates construction from industries that are oriented to more impersonal commodity markets. The central problem of construction is the form that relation takes.

As this interaction changes character, the ultimate product is altered as well. It may be argued that this applies to both historical epochs and the interval of time spanned by the building process. The relation between the consumer and supplier changes as the building moves from one stage of its life cycle to another. Their purposes evolve in time, as well. Preferences and expectations are continually revised during the building process as new knowledge is acquired. In the passage of time, the expectations of various economic agents may converge, but may also diverge and result in direct conflict.

PLANS AND PLAN REVISIONS

We are interested in buildings as capital goods that, together with other capital goods, form the capital combination underlying a production plan. As economic conditions change, production plans change, too. The revision of plans entails reshuffling capital combinations. It follows that at any moment some buildings and other durable capital goods are being used for purposes different from those envisaged when they were designed.[11]

By means of capital maintenance and replacement a capital combination may be adjusted to reflect changes in a production plan. Among other long-lived capital goods, buildings are continually transformed by a multitude of minor alterations. However, maintenance and replacement are usually seen restrictively as ways to preserve the value of an investment—essentially rearguard actions— rather than as responses to changing ideas in a larger economic arena. Consequently, capital maintenance and replacement are by and large neglected by economic theory [Lachmann, 1986: 67–69].[12]

Another aspect of capital utilization concerns the intensity of capital use per unit of time, that is, the rate of capital utilization. For example, a capital asset may be used in one, two, or even three shifts. The intensity of capital utilization will in many cases be different from that originally planned. As a consequence, the attendant maintenance and replacement expenditures may differ significantly from those envisaged in the process of design. Interest in this aspect of changing capital utilization has been relatively meager as well [Foss, 1984: xi].

As the building process unfolds, economic conditions are also changing. Under new conditions, an originally satisfactory building may require significant redesign and reconstruction. Such a decision may in turn affect the client's production plan, and require new investment for rebuilding. In this way, the building process interacts with evolving economic conditions. The interaction of business and building cycles is especially important in this context, as we will see in Chaps. 3 and 4.

The nature of the building stock of a country or an organization, and even the configuration of a particular building, limits the options for redesign, reconstruction, and reuse. Many building components, individual buildings, and entire city blocks reflect malinvestment in the past. Much of this wasted effort is a result of misgauging future economic conditions combined with the rigidity of capital resources.

As long-lived capital goods, buildings are heterogeneous and specific. Converting them to new uses is often very costly. However, rigidity characterizes not only completed buildings, but also the early phases of the building process, when a particular building is still far from being materialized. Every phase of the building process constrains subsequent decisions, but this is especially true of the early phases, when the ossification of a building's final form proceeds most quickly. Flexibility and adaptability provided in the early phases may facilitate conversion in the future. We will explore these issues in considerable detail in Chaps. 3 and 4, as well.

To understand why building processes fail, that is, why some buildings turn out to involve malinvestment of resources, we must first ask how they ever succeed.[13] Successful building processes require that all protagonists frame mutually consistent plans. In reality, it is unlikely that everyone's plans will be perfectly coordinated.

INVESTMENT AND INTEREST

We are interested here in buildings that are capital assets. A decision to construct such a building is an investment decision. In economics, this is the domain of capital budgeting and benefit-cost analysis.[14] It encompasses such topics as selection, timing, and amount of investment within any given period, as well as the arrangement of the financial means for the implementation of investment projects. It also encompasses disinvestment decisions. Discounting permeates this field.

Capitalization and annualization are two mutually consistent and symmetrical methods for making payments and receipts, or, more generally, inputs and outputs of an economic process, commensurable over time. We will focus on capitalization. Any payment or receipt can be moved to a later date by compounding it, that is, by adding interest to it, or to an earlier date by discounting it. If we capitalize in this fashion to the beginning of the process to obtain the present value, we must discount the latter payments or receipts according to the deferment. The interest or discount rate itself reflects intertemporal choices. As the interest rate increases, the future value of a payment or receipt increases, while the present value of a payment or receipt decreases as the discount rate increases. Therefore, an increase in the interest rate will tend to make future payments less attractive and future receipts more attractive than before, while the opposite will hold for an increase in the discount rate.

Interest was for centuries associated with usury.[15] In the Middle Ages, Thomas Aquinas wrote that interest was a "payment for time," and that no such payment could be justified since time was a gift of the Creator to which we all have a natural right. Interest actually stems from the need to make systematic trade-offs between the present and future allocations of scarce resources. The interest or discount rate can be thought of as an exchange rate between present and future values.

ECONOMIC TIME-HORIZON

It is often argued that the time-horizon of an economic agent—a firm, for example—should not extend too far into the future, because very distant inputs and outputs can generally be taken to be relatively unimportant.[16] This is explained in terms of basic discounting principles. Briefly, the money value of very distant inputs and outputs of the firm will be heavily discounted in calculating their present value. The results of calculations based on the infinite time-horizon thus differ little from those based on rather short time-horizons. For example, it can be shown that the present value of an annuity of one dollar over ten years at a discount rate of 10 percent is in excess of six dollars; its present value over 25 years is in excess of nine dollars; the present value of a perpetuity, other things remaining equal, is only ten dollars.

An alternative argument for a relatively short economic time-horizon concerns our limited ability to foresee the future consequences of our actions, an important aspect of our "bounded rationality." More generally, "[t]ime is a denial of the omnipotence of reason" [Shackle, 1972: 27, quoted by Lachmann, 1982: 37]. The longer the time-horizon, the more this argument applies. As Simon [1981: 179–80] argues,

> [T]he events and prospective events that enter into our value system are all dated, and the importance we attach to them generally drops off sharply with their distance in time. For the creatures of bounded rationality that we are, this is fortunate. If our decisions depended equally upon their remote and their proximate consequences, we could never act but would be forever lost in thought. By applying a heavy discount factor to events, attenuating them with their remoteness in time and space, we reduce our problems of choice to a size commensurate with our limited computing abilities.

This is reasonable, and yet it may induce a myopic perspective, which is especially misleading when we are considering long-lived capital assets, such as buildings. For instance, a trade-off between a dollar spent on a building and a computer will tend to be biased toward the computer, simply because the latter has an expected service life significantly closer to the adopted time-horizon. Furthermore, because costs tend to outweigh benefits in the early stages of long-lived investment projects, it follows that myopic investment choices would tend to curtail the volume of investment.[17] As the cut appears to be arbitrary, we are tacitly advised to reduce the time-horizon to a minimum. Paradoxically, the infinite time-horizon and myopia complement each other.

Of course, distant inputs and outputs may be quite significant in present value calculations: for a given interest or discount rate and a given date, the magnitude of an input or output is of primary importance. The greater this magnitude, the more likely it is that we will neglect it at our peril. This is hardly surprising, but it is nevertheless worth mentioning because it shifts our vision toward the future. This vision is indispensable to building professionals, especially architectural and engineering designers. Something approaching a geologist's conception of time is needed here, in which both gradual and abrupt changes play an important part in the overall picture.

TIME-PROFILE OF AN ECONOMIC PROCESS

How far should we go, however? The answer is conceptually simple: we should consider the life cycle of an economic process, including all the underlying building activity. All relevant inputs and outputs will thus be accounted for.

Naturally, this refers to the economic, not physical life cycle, as the only requirement concerning the physical durability of capital is that it extend beyond the economic horizon of an economic agent. For this purpose we need a model of an economic process.[18] Some aspects of this model will be useful to us throughout this book. The process may be regarded in two ways: *ex ante,* as a plan, and *ex post,* as historical accounts.[19] We will focus on the former.

Consider a firm that converts a flow of inputs into a flow of outputs. We can represent the anticipated variation of these flows over time by an input-output profile, shown in Fig. 1.1. The economic process consists of the construction of a plant, its operation over a period of time, and its ultimate dismantling. Here, "plant" refers to the capital combination underlying a production plan. The process is accordingly mortal: it will have a beginning and an end. The time-profile of inputs displays a "hump" associated with construction expenditures, while maintenance and replacement expenditures are included in the subsequent inputs. If plant reconstruction is anticipated, it can be represented analogously by another such hump, as shown in Fig. 1.2. In addition to the input-output profile, each process can be represented by a capital value profile, which shows the discounted value of the remainder of the process in each period. This is shown in Fig. 1.3. However, this is not a market value, but a value anticipated in planning.

The capital value profile, unlike the input-output profile, will depend on the interest rate, which is determined by macroeconomic forces, that is, given exogenously. A fall in the rate of interest will raise the capital value curve, because in that case the future net benefits will be discounted less, and vice versa. By the same token, a fall in the rate of interest will lengthen the economic life of a

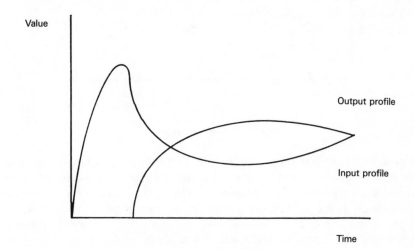

Figure 1.1 Input-output profile.

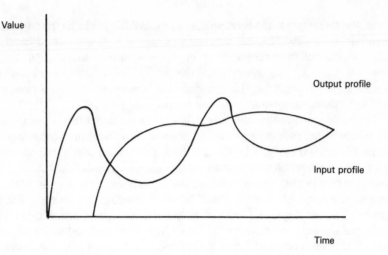

Figure 1.2   Input-output profile with plant reconstruction.

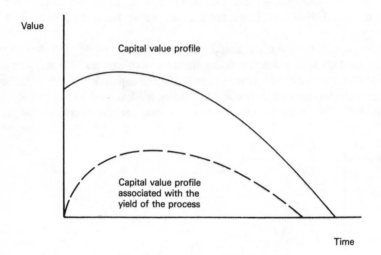

Figure 1.3   Capital value profile.

process, that is, shift the economic horizon to the right. A sufficient rise in the rate of interest will therefore reduce the capital value at the beginning of the process to zero, as shown by the lower curve in Fig. 1.3. This rate of interest can be identified as the yield of the process, or its internal rate of return. At this value we will be indifferent between investing and saving. At a rate of interest higher than the internal rate of return it will not be profitable to carry on the process at all, because in that case the capital value at the beginning of the process will be less than zero.

It should be mentioned in passing that interest rate shifts may occur at any time during the entire economic process. Although we are presently focusing on the anticipated shifts in the rate of interest, we should keep in mind that changes in economic conditions cannot be anticipated with certainty. A degree of speculation is therefore always involved at the beginning of an economic process. In fact, speculation ceases only upon termination of the process.

The obvious test of viability of an economic process is that it should yield at least the market rate of interest. The familiar discounting principles provide the foundations for the calculation. If we reckon inputs as negative and outputs as positive values, corresponding to the costs and benefits of the process, this means that the present value of the flow of net outputs must be nonnegative. This nonnegativity condition is necessary, but not sufficient. If the process went on forever, it would be a sufficient condition, as well. The decision to terminate the process must itself be based on economic considerations: it must be more profitable up to the economic horizon than to any other period. The economic horizon is the key variable here. The test of profitability is the present value calculation again; namely, the present value of the process must be greater for the economic horizon than for any other time-horizon.

Clearly, in considering the economic viability of such a productive process, the second hump—indicating the anticipated plant reconstruction depicted in Fig. 1.2—must also be included in the picture. It is crucial for our present purposes that the second hump is likely to be fairly significant in this calculation. At least, we cannot abstract it away because of some general rule concerning remoteness in time, or on any other *a priori* grounds. In short, the life cycle of the economic process should be considered as a whole, from its beginning to its end.

We should note that sunk costs are irrelevant in investment decisions. Each investment project must be considered on its own merit. The fact that present decisions cannot change the decisions made in the past reflects the irreversibility of time. This insight is particularly relevant to reconstruction, if it was not anticipated from the very beginning. In that case, the viability of reconstruction is calculated in precisely the same way as that of any new investment project.

BUILDING PROCESS AND ECONOMIC PROCESS

At this point the reader may ask why we are discussing inputs and outputs, rather than buildings, pure and simple? It is because buildings must be considered in the context of the economic process of which they are a part. First, the minimization of any component of building costs does not guarantee that total costs will be reduced, let alone minimized. For example, a reduction of construction costs may result in an increase of maintenance and replacement costs that offsets the initial savings. Second, and even more important, both costs and benefits should be taken into account. For instance, an increase of construction costs may result in an even greater increase of benefits associated with building utilization. The minimization of discounted value of the costs incurred over the building life cycle is justifiable only when we can assume that the benefits will be unaffected by changes in costs.

Building professionals can make meaningful proposals for changes in planned allocations only if they understand the relationship between the building and the remainder of the economic process. This applies to trade-offs relating to space and time, as well as their interdependence. In particular, the problem of matching building component maintenance and replacement cycles with the time-horizons of owner and/or user's needs becomes central. Among professionals concerned with buildings, it is the architectural designer who should be most knowledgeable about this process and the physical structure that "contains" it. The other professionals participating in planning, construction, and management of buildings cover relatively narrow domains of expertise.

Building economics offers a systematic framework for the representation of building processes in time. An economic evaluation of a building is in fact impossible without the time-profile of the underlying economic process. This can be perceived as a problem of finding some numbers to "plug into" the prescribed formulas, but it can also be perceived as a valuable framework for considering design, building, and management in time. The latter perspective is undeniably more challenging and stimulating.[20]

To be sure, this is not to say that architectural design should be subservient to economic purpose. The built environment already suffers from too many buildings designed and built with petty economics in mind. As Galbraith [1984: 61] states emphatically, "[t]he good architect must always be at odds with economics, including the economic constraints of his client." But to be at odds with economics does not mean to disregard it. On the contrary, to reach his or her objectives, the designer needs to understand and appreciate the economic process that motivates the client's behavior.[21]

REAL PROPERTY

In general, buildings are tied to a particular location over an extended interval of time. Immobility, that is, locational rigidity, is one of their essential characteristics. The notion of real property unifies buildings and land across both spatial and temporal dimensions.

Building economics treats parcels of land as it treats buildings, as interactive factors in an economic process. The building life cycle may be seen as part of the real property cycle, starting and ending with a vacant lot, the so-called unimproved land. The real property cycle is shown diagrammatically in Fig. 1.4. As the economic life of a building is intertwined with the economic value of the land it occupies, building economics partially overlaps the realm of land

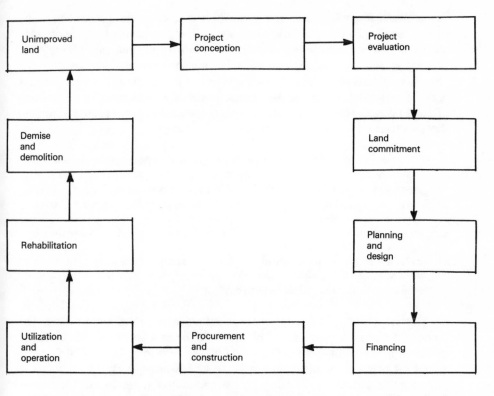

Figure 1.4 Real property cycle.

economics. Many a fine building has been allowed to deteriorate, torn down, and replaced by another building because the original one could not generate sufficient return in view of the value of land itself.

In some cases the real property cycle involves shortcuts and detours of various kinds. For example, building demolition may be a part of the construction phase, in which case all the phases from project conception to financing in fact precede demolition itself, and land commitment is skipped altogether. Of course, each phase of the cycle provides many opportunities for entrepreneurial error, delay, and change in plans of economic agents. Fig. 1.4 therefore portrays the "longest" real property cycle possible.

Because it is only a slight exaggeration to say that cities are aggregates of constructed facilities—buildings, roads, bridges, etc.—building economics also offers some insights into the microeconomic foundations of urban economics. From the vantage point of building economics, cities are spontaneous creations of innumerable individuals engaged in building processes culminating in distinct constructed facilities on separate parcels of land.

Although capital and land are distinct economic concepts, as evidenced by the fact that land—a permanent asset—is not depreciable, most so-called fixed capital is inherently linked to land use. Specific links are not permanent, even though they may span long intervals of time. The permanent character of land should not be confused with the temporary character of land use. Therefore, both capital and land use are characterized by nonpermanence.

Should land be included in the definition of capital goods? Hayek [1941] argues that it should not be, because nonpermanency is an essential characteristic of capital goods. Hayek's [1941: 51–52] conception of nonpermanency deserves special attention because of its subjectivist character:

> The main point to be kept in mind is that what matters is not permanency in any absolute sense, but the opinion of the economic subject as to whether particular resources will last throughout the period in which he is interested [. . .]. [. . .] What may be regarded as an even more fundamental basis for the distinction [between permanent and nonpermanent goods] is the fact that the future services of some resources cannot be anticipated [. . .]. The underlying fact, and in a sense the most general aspect of the phenomenon under consideration, is the *irreversibility of time* which puts the future services of certain resources beyond our reach in the present and so makes it impossible to anticipate their use, whereas the present services of those resources can as a rule be postponed.

Hayek [1941: 57] distinguishes his definition of capital from that of the "produced means of production," of which the latter is more inclusive. Those produced means of production that might be expected to remain useful forever would not be capital in his definition of the term. Consequently, he argues that most durable capital goods that could be understood as improvements to the land, such as buildings, are better understood as land. According to Wicksell [1934: 186], quoted by Hayek [1941: 57–58n]:

> Such improvements to the land often leave a permanent residual benefit. [. . .] Thus new qualities which, once acquired, the land retains for all posterity, cannot be distinguished either physically or economically from the original powers of the soil [. . .]. It may be further pointed out that nearly all the long-term capital investments, nearly all so-called fixed capital (houses, buildings, durable machinery, etc.) are, economically speaking, on the border-line between capital in the strict sense, and land.

Therefore, for Hayek and Wicksell, capital might not include buildings and other constructed facilities, let alone land. However, it is interesting to note in passing that the "borderline" status of fixed capital clearly points at the common ground shared by building economics and land economics.

Now, Lachmann [1976a: 147–48] explicitly includes both land and buildings in his definition of capital. He is consistent in this regard throughout his writings. Although Lachmann [1978a] follows Hayek [1941] in the essentials of his theory of capital, he disagrees with Hayek's position on land (and fixed assets). Lachmann [1978a: 11] includes land in the definition of capital goods on the basis of its use in the production process:

> Our interest lies in the uses to which a resource is put. In this respect land is no different from other resources. Every capital combination is in fact a combination of land and other resources. Changes in the composition of such combinations are of just as much interest to us where land is, as where land is not affected.

Furthermore, Lachmann [1978a: 12] explicitly dismisses Hayek's criterion of nonpermanence, on the ground that it is irrelevant from the point of view of the uses to which resources are put. Lachmann is concerned with capital combinations, not capital as such.

Lachmann is correct insofar as he is dealing with *land use* and not with land *sensu stricto*. Land use is obviously not permanent. Also, land is often rented or leased for a definite period of time and for a limited use (or set of compatible uses). However, it is important to keep in mind the subjectivist aspect of nonpermanence discussed by Hayek. The anticipation of future land use is therefore central in this case, as well. It is reasonable to argue that land use patterns are perceived by economic agents as increasingly ephemeral because economic change has become a dominant aspect of their expectations about the future.

CONCLUSION

Our objective is to explain some of the economic forces influencing patterns of investment in and uses of buildings. This perspective should aid decision making for both clients and professionals during the building process. An enhanced understanding of how economic forces shape individual buildings, clusters of buildings used for a common purpose, and the built environment will, we hope, contribute to better collaboration in the future.

We intend to lay some of the groundwork needed before a coherent theory can be attempted. As a special theory of capital, building economics should contribute to economic theory, not just passively apply economic concepts and principles to a new area of study. In the long run, building economics can develop and thrive only if it is fruitful in this sense.

One of the key propositions introduced in this chapter is that building economy cannot be meaningfully discussed outside the framework of the entire economic process of which the building is a part. In particular, we have seen that building activity cannot be properly understood without an understanding of the production plan that underlies it. As this plan is subject to continual revision in view of changing preferences and expectations, so is the building itself. It must be adaptable to changing economic conditions if it is not to be replaced, disposed of, or simply abandoned.

Before we proceed with an economic interpretation of the building process, as well as the interaction between building and business cycles, we will focus on some fundamental concepts and propositions of the theory of capital. Throughout, we will concentrate on issues of specific interest to building economics, as a special theory of capital, rather than on the theory of capital in its most general form. Nonetheless, some general considerations will be unavoidable. It should be kept in mind that the concepts discussed will be invaluable for a full understanding of the research prospects of building economics, addressed in the concluding chapter.

NOTES

1. As Mises [1966: 20] argues, "[w]hen applied to the means chosen for the attainment of ends, the terms rational and irrational imply a judgment about the expediency and adequacy of the procedure employed." For Mises [1966: 21], however, the ends themselves should be regarded as given. What is economic is only the conduct of people, rather than their purposes [Mises, 1976: 61]. These purposes reflect the subjective preferences of economic agents. For an extensive discussion of this position and its implications, see Lachmann [1982].

2. As Hayek [1941: 3] writes, the main concern of the theory of capital is "to discuss in general terms what type of equipment it will be most profitable to create under various conditions, and how the equipment existing at any moment will be used, rather than to explain the factors that determined the value of a given stock of productive equipment and of the income that will be derived from it." Lachmann [1978a: vii and 1986: 59–60] thus argues that it may be useful to distinguish between the *theory of capital* and *capital theory* along the lines of Hayek's distinction. The former concern is intrinsic to the concept of capital, while the latter is subservient to the theory of interest.

3. According to Samuelson [1967: 570], "[c]apital theory is one of the most difficult parts of economic theory." Solow [1963: 9], for example, writes:

   > The theory of capital has been a subject of controversy among economists at least since Torrens attacked Ricardo's theory of value in 1818. [. . .] But the status of capital theory is still unsettled.

   As Lachmann [1978a: 1] argues:

   > The realm of economics consists of many provinces between which, in the course of time, a fairly high degree of interregional division of labor has evolved. Naturally, development in some of these regions has been faster than in others. There are some "backward areas," and a few of them actually appear to merit description as "distressed areas." None seems to have a better claim to this unenviable status than the Theory of Capital. In fact it would hardly be an exaggeration to say that at the present time a systematic theory of capital scarcely exists.

4. The most famous polemics are those between Böhm-Bawerk and Clark in the 1890s and between Hayek and Knight in 1930s, but Austrian economists also made a significant contribution to the controversy between capital theorists in the two Cambridges in 1960s and 1970s. For a brief historical overview of the theory of capital from Adam Smith to Keynes, including Böhm-Bawerk and Clark, see Kregel [1976: Chap. 2]. For a discussion of the lasting contributions of the Austrian school, stemming in part from the polemics on capital, see Hicks and Weber [1973], for example.

5. According to Mises [1966: 1-2]:

> The economization of time is independent of the economization of economic goods and services. Even in the land of Cockaigne man would be forced to economize time, provided he were not immortal and not endowed with eternal youth and indestructible health and vigor. Although all his appetites could be satisfied immediately without any expenditure of labor, he would have to arrange his time schedule, as there are states of satisfaction which are incompatible and cannot be consummated at the same time. For this man, too, time would be scarce and subject to the aspect of *sooner* and *later*.

6. Of course, in some cases the building client and the user of building services are not necessarily the same. This distinction is not crucial here. However, in some instances discussed in this book, the distinction between clients and users, as well as various parts of the client's organization, will be unavoidable. The corporate building client is nowadays a complex entity requiring coordination during the building process. The architectural designer is thus often forced to play the role of an unofficial coordinator. As Nisbet [1961: 18] observes, sometimes this is a rather difficult task in view of the "schizophrenic character of the corporate client."

7. On the notion that a construction project represents a temporary market, see Hillebrandt [1974: 161].

8. See, for example, Turin [1966], Stone [1983, first published in 1966], Cassimatis [1969], Hillebrandt [1974 and 1984], Lange and Mills [1979], Dolan [1979], Goldthwaite [1980], Powell [1980], and Shutt [1982]. Although most of these writers focus on construction issues, they of course discuss other phases and facets of the building process. Note that some of the references cited—most notably those by Turin and Stone—predate by a decade the emergence of building economics as a bona fide field.

9. As Collier and Halperin [1984: 3] write, with specific reference to real estate developers:

> It all begins in the mind's eye. Amid the inner recesses of the creative mind of a venturesome developer, a few wisps of an idea begin to come together. The idea takes shape first as a promising concept, then, gathering substance, it grows and finally springs forth as a full blown plan for a development project.

10. More generally, in Mises' [1966: 92] words, "[e]conomics is not about things and tangible material objects; it is about men, their meanings and actions." This distinction should be borne in mind throughout this book.

11. As Lachmann [1978a: 3] argues:

> Each capital good is, at every moment, devoted to what in the circumstances appears to its owner to be its "best," i.e., its most profitable use. The word "best" indicates a position on a scale of alternative possibilities. Changing circumstances will change that position. [. . .] Hence, we cannot be surprised to find that at each moment some durable capital goods are not being used for the purpose for which they were originally designed. [. . .] In each case the change in use means that the original plan in which the capital good was meant to play its part has gone astray. In most of the arguments about capital encountered today these facts and their implications, many of them crucial to a clear understanding of the nature of economic progress, are almost completely ignored.

12. However, Lachmann [1986: 68] notes that today the business community is conversant with the practical relevance of capital maintenance and replacement. For example, Bromiley [1986: 41] recently interviewed a planner from an American corporation who said that "one of the areas that get hit first when things start slowing down is maintenance." Bromiley [1986: 53] also quotes a senior divisional planner from another American corporation concerning capital replacement:

> From the plant managers we get a set of proposals on what they need to spend to keep the plants going. A lot of this is replacement and minor changes.

Surprisingly, Bromiley's book, concerning the question of "[w]hat determines the level of corporate expenditures on property, plant, and equipment" [Bromiley, 1986: 1], does not contain a single reference to maintenance and replacement as theoretically significant categories. Of course, this is largely due to the very problem addressed—that of the *level* of investment, as distinct from the *pattern* of investment and capital use stressed by the Austrian economists. (See note 2.)

13. On this methodological point, see Hayek [1948: 34].

14. For a succinct overview of capital budgeting, together with the fundamentals of benefit-cost analysis, see Baumol [1977: Chap. 25], for example. Baumol also introduces the discounting principles used below.

15. See for example, Böhm-Bawerk [1959a: Chap. 2], Fisher [1965: Chap. 3], and Goldthwaite [1980: Chap. 2].

16. See Baumol [1970: 64, 67], for example.

17. Hayes and Garvin [1982] go so far as to claim that the project evaluation techniques based on discounting are responsible for the slowdown of the growth of capital investment in the United States. They submit that the discounting approach is "biased against investment because of critical errors in the way theory is applied" [Hayes and Garvin, 1982: 71–72]. The excessive "premiums" and "contingencies" built into the short payback periods and high hurdle rates of return imposed upon investment projects are examples of this. In response to these authors, Bierman and Smidt [1984: 7] argue that "[t]here is no inherent bias in the tool." However, it is of course essential to use it properly. (For further discussion, see Bon [1986a].)

18. This model is based on Hicks' [1973: Chap. 2] "neo-Austrian" model, which is in many respects related to Hicks' [1946 and 1965] earlier models. (For a succinct presentation of the earlier models, see Baumol [1970: Chap. 5], for example.) Figures 1.1 and 1.3 also come from Hicks [1973: 15, 19]. Note that all the flows are rendered homogeneous by considering their money value, given that input and output flows are reckoned as negative and positive values, respectively. This homogeneity assumption, together with the underlying assumption of stationary equilibrium, will be abandoned in Chap. 2, however.

19. The distinction between *ex ante* and *ex post* views of an economic process, that is, between prospective and retrospective calculations of economic quantities, was introduced by Myrdal [1939]. For a discussion of this important contribution, see Shackle [1967: Chap. 10]. This distinction will be encountered frequently in the chapters that follow.

20. For an elaboration of this argument, see Bon [1986c].

21. It is perhaps worth emphasizing the difference between the conception of the building process advocated here (where it is understood as a part of an individual economic process, itself an "atom" of the production process as a whole) and the conception of the individual building process as the atom of analysis (where the building industry is defined in terms of aggregations of such atoms). The former conception ultimately concerns the entire production sphere, whereas the latter focuses on the building industry. For a discussion of the latter view of the building process, based on the work of Turin [1966], see Groàk [1983], for example.

2

Theory of Capital

INTRODUCTION

In Chap. 1, we introduced many subjects and concepts that require more careful scrutiny. The basic propositions of the theory of capital will be elucidated in this chapter. Before we proceed to the discussion of the building process, we ought to step back and consider the interconnections between the key concepts that link building economics and the theory of capital.

Some readers may find this chapter unduly abstract. They may also find it expendable and be tempted to skip it altogether. Although this chapter demands more from the reader than others, we contend that some travail at this stage is worthwhile for the full comprehension of the chapters that follow.

We will first analyze the nature of human needs, including those for shelter. For this purpose, a conjectural history of building activity will be presented. A classification of all goods into consumption and production goods will be introduced here. This classification will be useful in explaining the nature of economic value, including the value of buildings as long-lived capital goods. Second, we will examine the connection between the concept of capital and durability of goods. We will show that durability is not a necessary characteristic of capital goods. Third, economic choices and plans made by individuals will be examined in the context of the market process. The notion of production plans will be discussed in some detail to facilitate the understanding of the role of buildings in the production process. Fourth, an overview of the Austrian theory of capital will be provided, followed by a discussion of capital heterogeneity, convertibility, complementarity, and substitutability. These concepts will be essential throughout this book. Fifth, we will introduce a classification of capital goods, which will be useful in the discussion of production plans and asset portfolios. Again, much of this book will remain incomprehensible without these concepts. Sixth, we will return to the value of complementary capital goods in general, and buildings in particular. Seventh, the problem of monetary calculation and capital accounting will be briefly discussed. We will show that the value of capital goods always reflects the subjective judgement of economic agents about future economic conditions.

The reader should be warned at this juncture that the presentation of many ideas here and elsewhere in this book does not correspond to their historical development. Those familiar with economic theory may occasionally feel that the superimposition of economic concepts developed at different historical periods under specific social conditions is outright anachronistic. However, it should be remembered that the objective of this book is to assemble in one place those concepts that may contribute to the development of building economics as a distinct discipline, rather than to present all the useful economic concepts in a way that is relevant for their exposition in terms of economic theory as a whole.

HUMAN NEEDS, SHELTER, AND BUILDING

Shelter is one of the foremost human needs. At first, shelter was appropriated in its natural form, without human alteration. Cave dwellings are the classic example. Like food, shelter was provided directly by nature. The only form of economizing at the dawn of civilization concerned the juxtaposition of satisfactory shelter and areas where food could be gathered in sufficient quantity.

The advent of animal husbandry and nomadic wandering in search of better pasture introduced a need for temporary shelter for both people and animals. The art of building has its roots in these rudimentary structures. Protection from a predominantly hostile environment probably first took the form of fenced enclosures; enclosure from above gradually followed. Economizing the use of scarce resources could already be divided into two components: economizing the use of time available for various activities, including building, and economizing the use of other resources that could be used for various purposes, again including building. When, where, with what means, and how to build required rational and purposeful decisions even in the case of these temporary structures.

With the development of agriculture, the first permanent buildings started taking shape. The land needed for buildings was appropriated directly from nature, as was the land needed for agricultural production. Buildings were intended to shelter not only people and animals, but also agricultural produce. While a portion of the produce was consumed, a portion had to be saved for another season. The buildings were needed to protect the seed, representing prospective food, over at least one season.

The needs for food and shelter were at first intertwined. Gradually, buildings were differentiated in terms of the needs they were intended to serve: some directly served the need for human shelter, while others served the need to preserve the food, animals, and the emerging agricultural implements, which in turn directly or indirectly satisfied human needs.

We should note in passing that no individual or single group of people invented buildings or human settlements formed by clusters of buildings. They arose through spontaneous human activity and evolved over extended periods of time. As economies progressed from production for direct consumption, to production on order, to production for the market, clusters of buildings and communication links between them evolved into an increasingly complex physical infrastructure underlying the economic process. However, it is worth emphasizing that human settlements were formed as a result of building decisions made by individuals, who more or less independently decided on the best way to satisfy their needs for shelter.

Now, all goods can be classified in a conceptually useful way into orders according to the distance, in terms of production stages, from the direct satisfaction of human needs.[1] As human shelter, buildings are first order goods, just like foodstuffs. As shelter for these foodstuffs, buildings are second order goods.

They satisfy human needs indirectly. Furthermore, as shelter for draft animals and agricultural implements needed for the production of food, buildings are third order goods. They are even farther removed from the satisfaction of primary human needs. In the course of civilization, as the number of production stages has increased, this progression has been extended to buildings as ever higher order goods.

Consequently, buildings serve purposes ranging from the satisfaction of the most immediate needs to those progressively removed in time and space. Buildings that fall in the former category are consumption goods, while those that fall in the latter are production goods. In this context it should be noted that building materials and tools are goods one order higher than the buildings they are used to construct. The order of building materials and tools thus depends on the order of buildings themselves. Of course, the classification of goods into orders does not concern their technical characteristics, but rather the economic purposes associated with these goods. This classification is obviously not intended to classify goods as objects of economic activity, but to relate them to the purposes of human beings, the subjects of economic activity.

Compared to other goods, buildings became objects of exchange rather late in human development. This is true primarily because buildings are difficult or impossible to move from place to place, and because they are connected to land, a resource which itself entered the sphere of economic exchange late in the development of civilization. Buildings and land are considerably less marketable, that is, less readily exchangeable, than numerous other goods, the most marketable of which gradually took the role of money—the medium of exchange. However, building services most likely started being exchanged for other services and ultimately for money much earlier than buildings themselves. Together with the gradual division of labor, building activity was progressively differentiated from other useful activities performed singly or collectively. The art of building became specialized as did many other useful activities that required skill and special knowledge acquired through experience.

This endeavor was also gradually differentiated as an ever larger number of building types evolved. Each building type was associated with a specialized building technique, requiring a particular quantity and quality of building resources. The first elements of building design probably arose with the need to combine the basic building types already mastered in previous stages of development. If a larger building were needed, a particular building type was repeated several times. If a multifunctional building were required, several existing building types were combined to form a single structure. A rudimentary form of designing arose as it became necessary to merge and join the building types already tested by previous experience.

As more complex buildings were needed, designing gradually emerged as an art separate from building itself and entered the sphere of exchange as a distinct activity. Because buildings and land are immobile, and as long as the level of demand for building services remained low, builders and designers moved from site to site as itinerant craftsmen covering wide territories in search of clients.

As markets for buildings and building-related services developed, these goods and services began to be exchanged for money, the universal exchange medium. In this way, their value became increasingly intertwined with the value of all other goods and services. The value of buildings and building-related services ultimately derives from the value people attach to the satisfaction of their immediate needs. In other words, the value of higher order goods is determined by the value of lower order goods, whose value is ultimately determined by that of first order goods.[2]

THE CONCEPT OF CAPITAL AND DURABILITY OF GOODS

The classification of goods into those of first and higher orders, that is, consumption and production goods, does not coincide with a classification of wealth into capital and noncapital. The term "capital" is often used to refer to all items of wealth that yield a permanent income. A consistent extension of this concept of wealth would include labor power, land, and all consumption goods of any durability in the concept of capital [Menger, 1981: 303].

A mistake frequently made in defining capital and classifying capital goods is an emphasis on the technical, rather than economic, perspective. For example, stress is often laid on the durability of goods regardless of their economic character, as can be seen in the Appendix.[3] However, capital consists only of those quantities of economic goods that are available now for use in the future [Menger, 1981: 303]. More specifically, durable consumption goods can directly satisfy human needs both now and later, whereas durable production goods can indirectly satisfy them only in the future. The concept of capital is therefore inherently related to that of expectations of the future, as will be argued in the sections that follow.

Two conditions must be met simultaneously for an economic good to be considered a capital good. First, an economizing individual must control it long enough for a production process to take place. Second, the same individual must also have either direct or indirect command of the complementary goods of higher order required to produce goods of lower order. In terms of real property, for example, direct and indirect command refer to owning and leasing, respectively, of buildings and land. A capital combination is a particular constellation of complementary production goods needed for production of a definite quantity of a given production good of lower order, or a given consumption good. Capital thus involves a combination of economic goods of higher order. As Menger [1981: 304] writes:

> The most important difference between capital on the one hand and items of wealth that yield an income (land, buildings, etc.) on the other is that the latter are *concrete*

durable goods whose services themselves have both goods character and economic character, whereas capital represents, directly or indirectly, a *combination* of economic goods of higher order (i.e., complementary quantities of these goods) whose services also have economic character and therefore yield income, but whose productivity is of an essentially different nature than that of durable wealth that is not capital.

Menger [1981: 305] thus concludes that "[a]lmost all difficulties that have arisen in the theory of capital can be traced to the linguistic confusion involved in including both of the above sources of income in the concept of capital."

The difficulties with the concept of capital are compounded by the ordinary usage of the term "capital," which is generally interpreted as a sum of money. Although money facilitates the transfer of capital goods and the transformation of capital into any form desired, the concept of money should be differentiated from the concept of capital [Menger, 1981: 305].

Therefore, it is paramount that different forms of income, such as rent, interest, and profit, be clearly distinguished in defining capital, as well as in classifying capital goods. In particular, all items of wealth that can be classified as durable goods yielding a permanent income should be considered in the context of rent. Wealth in the form of money that yields a permanent income should be analyzed in the context of interest. Finally, the productive use of a capital combination in the production process, which yields an income through the sale of the production goods of lower order or the consumption goods produced, should be treated in the context of profit. As we will see shortly, a proper understanding of the connection between capital and profit is impossible without the concept of entrepreneurship.

Now, buildings are undoubtedly durable goods. But it is important to emphasize that a particular building may be classified as a durable consumption good providing direct satisfaction of its owner's needs; as a durable item of wealth yielding permanent income in the form of rent; or as a durable production good yielding income in the form of profit. In this book we are primarily interested in buildings as production goods. As a production good, a building forms a part of a capital combination used to produce lower order goods, which ultimately contribute to the production of consumption goods for the satisfaction of direct human needs. Only in such a capacity can a building be understood as a capital good. Put differently, durability of buildings plays no role in their classification as capital goods. Technical characteristics of buildings are irrelevant in this context also. Many a residential building has been transformed into an office building, a factory, or a warehouse. Similarly, many commercial office buildings, representing durable items of wealth from the vantage point of their owners, have been rented or leased to firms engaged in production of goods and services. What matters here is the role a building is expected to play in the production process.

INDIVIDUAL CHOICE AND THE MARKET PROCESS

Acts of individual choice are the fundamental building blocks of an economic process. Let us explore individual decisions in the context of the market process.[4] Individuals in the market economy pursue their purposes to the extent permitted by the resources available to them and the opportunities offered by these resources for the satisfaction of their needs. These opportunities are determined by each individual's knowledge of relevant technological possibilities, and of opportunities for exchange in the market for either the resources or for products. Each individual exploits both types of opportunities to the extent that he or she is aware of them. The exchange opportunities are at the same time determined by the choices other market participants are making in seeking opportunities to fulfill their own purposes. The market process involves a systematic and continual revision and adjustment of each individual's decisions in light of new knowledge gained from interaction with other market participants.[5]

The focus on learning and adjustment, as parts of the market process, is worth emphasizing. The choices of all the individuals interacting in the market are brought into consistency with each other *via* the market process, that is, by means of continual adjustment based on learning. However, learning and adjustment are not intrinsic only to a market process involving many participants. An isolated individual may also need to adjust his or her plan, based on learning. Robinson Crusoe represents such an isolated individual. We will return shortly to this conceptually useful distinction.

Because both consumption and production take time, the alternative consumption and production opportunities considered by an individual consist of possible courses of action over an interval of time. This involves multiperiod action. The individual's decision making consequently requires multiperiod planning. For example, the first period may involve building a production facility; the second period, acquisition of equipment; and the third period, production of particular goods. The essence of such a plan is the consistency of the activities it embraces. In a multiperiod plan, current activities are made consistent with future activities envisaged by the individual.

Clearly, an individual will be aware that his or her multiperiod plan may have to be changed or scrapped at a later date, and a new plan adopted. A multiperiod plan involves the individual's expectations about both the technological effects of his or her actions and the exchange opportunities that will arise in the future. The degree of uncertainty with which an individual is able to forecast the future will affect his or her multiperiod plan. The greater the uncertainty faced by the individual, the greater will be the flexibility of the adopted plan, that is, the degree to which irrevocable decisions will be reserved for later stages in the fulfillment of the plan.

An individual may appraise the adopted plan at three distinct points in time: before the implementation of the plan *(ex ante)*, after its completion *(ex post)*, or at some intermediate point in time. The appraisal will involve alternative multiperiod plans contemplated by the individual. Looking forward, the individual may favor a plan with initial stages identical to those required by several alternative plans, so that he or she is not committed at the outset to an irrevocable plan. Looking backward, an individual will attempt to gauge whether the completed plan was more or less successful in terms of gains or losses, than the rejected alternatives. However, it should be noted that such retrospective comparisons are inherently subjective, because the rejected alternatives were never realized. As we will see in Chap. 3, this view of foregone opportunities is fundamental to the understanding of opportunity costs.

Someone in a market economy enjoys considerably more freedom in his or her plans than an isolated individual. A person operating in isolation must consider only the resources that are or will be at his or her direct disposal. An individual in a market economy may consider the resources obtainable through exchange, including borrowing. Borrowing can be thought of as a form of intertemporal exchange. Furthermore, an isolated individual needs to produce only the consumption goods required to satisfy his or her needs, while an individual in a market economy may produce consumption goods to meet the needs of other market participants.

In a market economy, an individual's multiperiod plan will take into consideration the market prices of all the resources needed to execute the plan. These prices will make possible the calculation of all the costs associated with the plan. In other words, in a market economy the individual's plans will involve monetary calculation to facilitate the process of planning itself. Most important, monetary calculation will facilitate the appraisal of a multiperiod plan in terms of gains and losses associated with its implementation. For us it is particularly interesting that the system of prices makes it possible to evaluate in money terms the relationship between the stock of production goods at the disposal of an individual, and the value of consumption goods he or she may produce under alternative multiperiod plans.

It is important to emphasize that a market economy is not composed of isolated individuals who perform their calculations on the basis of market data. In fact, a market consists of many interlocking plans that directly influence one another. The market encourages coordination among the multiperiod plans. However, this coordination can never be perfect. A perfectly interlocking system of multiperiod plans would require all present plans to be mutually consistent not only in the present, but also in the future. Furthermore, it would imply mutual consistency of all the plans ever made, in the past or in the future.

An isolated individual or an individual in a market economy may decide to revise their plans as new knowledge is acquired during plan implementation. However, in a market economy plan revision may be required due to entrepreneurial error in anticipating the need for goods produced. An isolated individual cannot misjudge his or her ultimate needs for consumption goods in the same way. Although conditions may change, at some point the plan adopted by an isolated individual corresponded to his or her anticipated needs. In a market economy, however, a plan may be misguided from the very beginning. The passage of time changes nothing, except that it makes the individual increasingly aware of his or her planning error. This distinction between *ex ante* and *ex post* errors will be useful in Chap. 4.

In a market system based on multiperiod plans, many production goods designed for one purpose will eventually be either scrapped because of plan failure, or used for purposes not originally planned. The market process consists precisely in enforcing such adjustments. The multiperiod view outlined here emphasizes the possibility of reshuffling production goods following plan revision.[6] It should also be noted that the market process may correct the entrepreneurial errors made in the past through the reuse of production goods originally intended for different purposes.

For example, a production good few people want would tend to command a lower price than originally envisaged. This, in turn, might make it more attractive to some unanticipated buyers who may see a use for it other than the one the entrepreneur had planned. A speculatively built office building may have to be leased or sold at a price significantly lower than that expected. As a consequence, it may attract lessees or buyers who have not originally expected to find a building at such a low price, and who may use it for warehousing purposes. The essence of this adjustment process is entrepreneurial discovery of opportunities such as these.

Now, what is the role of the theory of capital in this context? First, a theory is required to explain why a particular array of capital goods is available at a given date, referring to the types of capital goods, the degrees of durability of capital goods, and so forth. Second, a theory of capital can explain the subsequent course of adjustment of capital goods. Plan revisions cannot be understood without reference to the stock of capital goods initially available. Clearly, buildings and other constructed facilities form a significant proportion of that stock. This view of the theory of capital is very different from those that describe it as an adjunct to the theory of interest.[7]

THEORY OF CAPITAL: AN OVERVIEW

We may conceive of the capital structure of an economy as an aggregate of capital combinations used by individual producers.[8] These capital combinations are, in turn, aggregates of heterogeneous capital goods that cannot be represented by any summary measure, or assessed "objectively" by an outside observer, but that nevertheless can be classified in a theoretically useful way, as will be shown shortly. The primary objective of the theory of capital is to explain how capital goods are used and how changes in their use occur in the market process.

The theory of capital must start at the level of individual action, which is based on a plan. In a production plan, capital goods and other resources are combined to produce desired outputs. These capital goods are complementary to one another within the plan. However, their individual contributions to the fulfillment of the plan may not be distinguishable. As parts of the capital combination, they contribute jointly to plan fulfillment.

Each production plan involves both operating assets and financial assets. Buildings are part of operating assets, as is the equipment housed in them. Financial assets or securities serve as instruments for obtaining the required physical assets. They represent the claims on the existing operating assets, and are traded in the stock exchange. The value of financial assets reflects the value of the physical assets, as well as the degree of planning success. The perception of planning success in turn depends on the expectations of the holders of financial assets.

As long as the plan is being fulfilled, the capital combination underlying it will be maintained. Whether the plan succeeds beyond the planner's expectation, or fails, the plan will be revised. The capital combination will be altered, and the capital structure as a whole will therefore change.

The success or failure of one production plan depends on the fulfillment of other individual plans in the market. The plans are interdependent. Only if all plans are consistent with one another will they all be fulfilled. In a market economy, however, it is unlikely that all individual production plans will be mutually consistent. Plan inconsistency implies plan failure, leading to plan revision. Finally, plan revision precipitates capital regrouping. Plan revision is thus the direct cause of changes in the capital structure.

Plans are based on planners' expectations. Unexpected changes leading to plan revision also require planners to change their expectations. Some changes are expected, or at least the variation in the underlying data does not surpass the normally observed variation attributable to "random" causes. Such changes will not result in plan revisions. Only the changes a planner believes to be due to "permanent" causes change his or her expectations and lead to a plan revision.

In a plan revision, some capital goods will be substituted for others. While complementarity is an aspect of a production plan, substitutability is an aspect of plan revision. These two concepts characterize respectively the coherence and adaptability of the capital structure. However, as we will argue shortly, a rigid distinction between these two concepts is unwarranted.

Insofar as plans depend on each other for their success, the failure of one plan may set in motion a series of plan failures. Since plans embody specific capital goods, some of which may be very durable and impossible to rapidly depreciate, plan failures imply capital losses. Buildings and other long-lived capital goods are examples of this. In some cases capital regrouping precipitates a cumulative failure throughout the economy. In other cases, when planners exceed their goals and experience capital gains, there may be a reverse cumulative process. Of course, capital gains and losses may also offset each other.

The uncertainty inherent in the capital formation process may lead to a chance clustering of errors. However, these errors are not systematically biased in either direction. Just as plan consistency throughout the economy is extremely unlikely, there is no reason to expect that plan inconsistency is biased either toward capital gains or losses. Cyclical swings may be an inevitable and unpredictable feature of market economies; however, it is possible that economic fluctuations have their origin outside the market process itself. We will return to this subject in Chap. 4.

The logic of interacting production plans informs the logic of the capital structure as a whole. At the firm level, the entrepreneur establishes the plan structure. It is modified by the market for the firm's securities, that is, the control structure. At the market level, the market process establishes the capital structure through the interacting network of plans. The individuals who made successful plans are rewarded, while those who made unsuccessful ones are punished. By the same token, the entrepreneurs who will determine the plan structure in the future are selected by the market.

Our presentation of the building process in Chap. 3 will rely on these concepts. The central notion in the theory of the building process is the interaction among the plans of all parties involved: building owner, designer, builder, user, and so forth. Again, there is no reason to expect that all the plans concerning a particular building project will be consistent with each other. The network of plans underlying the building process may fail either in part or *in toto*. This perspective on the interaction among the plans, derived from the theory of capital, will be used to elucidate various problems that may arise in the building process.

INVESTMENT AND CAPITAL HETEROGENEITY

Investment is often defined as addition to the capital stock. A realistic discussion of the incentive to invest cannot ignore existing capital resources, that is, the effect that investment will have on the use and profitability of the existing capital. This depends on whether we conceive of capital as homogeneous or heterogeneous.

A theory of investment based on the homogeneity assumption can deal only with quantitative capital change—investment and disinvestment, that is, accumulation and decumulation. It cannot deal with changes in the *composition* of capital goods. As long as we conceive of all capital as homogeneous, we *ipso facto* must regard all elements of a homogeneous aggregate as perfect substitutes for each other. We have to conclude that new capital must compete directly with the old and make it less profitable. The heterogeneity assumption changes this picture. Heterogeneity implies complementarity of capital goods.[9]

In this context it is important to introduce a distinction between concepts of stock and fund, that is, flow and service [Georgescu-Roegen, 1971: 220–28]. This distinction will help us understand the difficulties with the notion of quantitative capital change, which is inherent in the homogeneity assumption. In particular, the homogeneity assumption may be applicable to stocks, but not to funds of services. On the one hand, a stock of circulating capital—fuel and raw materials, for example—can be accumulated and decumulated by means of a flow. On the other hand, a fund of fixed capital—plant and equipment, for instance—can be used or wasted; it cannot be accumulated or decumulated, but it either does or does not provide services over an interval of time. As was argued in Chap. 1, a stock can be consumed at once, while a fund provides services over a period of time. The difference between the two concepts is therefore fundamental. In Georgescu-Roegen's [1971: 227–28] words:

> The difference between flow and service is so fundamental that it separates even the dimensionality of the two concepts. [...] The amount of flow is expressed in units appropriate to substances [...]. The rate of flow, on the other hand, has a mixed dimensionality, (substance)/(time). The situation is entirely reversed in the case of services. The amount of service has a mixed dimensionality in which time enters as a factor, (substance)x(time). [...] The rate of service is simply the size of the fund that provides the service and consequently is expressed in elemental units in which the time factor does not intervene. A rate with respect to time that is independent of time is, no doubt, a curiosity.

Let us return to the distinction between heterogeneity and homogeneity of capital goods. We will not be able to understand the investment pattern if we adopt the homogeneity hypothesis. Because the incentive to invest will depend on the expected effect of new capital on the earning capacity of old capital, investment decisions will depend on the composition of the existing capital stock. In general, investments will tend to be complementary to the existing capital [Lachmann, 1978a: 7]. In both magnitude and concrete form, investment will reflect the possibilities left open by the existing composition of the capital stock.

Economic progress involves reshuffling capital combinations, and maintaining and replacing capital goods. This recognition is a big step toward realizing that all progress does not involve investment in new plant and equipment, that is, new ventures; it often involves thousands of entrepreneurs continually reshuffling their existing capital combinations [Lachmann, 1977: 234].[10] In this context, new investment will also play a key role in reshuffling existing capital goods.

This view of investment is of course especially important for the so-called fixed capital goods, including buildings. Fixed capital goods typically have several potential uses. They also can be adapted to new uses by means of investment. Though they may be adapted to new uses, the concrete forms of existing capital goods cannot be abstracted away without losing sight of the very problem at hand. In other words, the homogeneity assumption is especially inappropriate when dealing with fixed capital goods, such as buildings. The provision of building services requires duration. This cannot be compressed in time beyond some limit expressed in terms of the rate of capital utilization. For example, plant and equipment cannot be used in more than a limited number of shifts per day. The perspective adopted here will be useful in considering the incentive to invest in buildings vis-a-vis other capital goods, discussed in Chap. 4.

CONVERTIBILITY OF CAPITAL GOODS

Capital goods are physically heterogeneous—they always have a definite form. There is no capital as such, in an abstract or ideal form [Mises, 1966: 503]. Similarly, there are no "generic" buildings. Capital goods defy aggregation primarily because of the diversity of purposes they serve. Therefore, a summary measure of a collection of physical capital goods is of no use to acting [Kirzner, 1976b: 138, following Mises, 1966]. This applies in different degrees to both fixed and circulating capital. The very distinction between fixed and circulating capital is one of degree, as well [Mises, 1966: 503].[11] However, this distinction may be made useful for heuristic purposes.

Capital goods have a more or less specific character. The degree of specificity is directly related to the process of production. The closer we come to the production of consumer goods, the more specific are the capital goods employed in the process of production. Furthermore, the more specific the capital goods, the less convertible will they tend to be.[12]

In other words, the notion of convertibility of capital goods is central to an understanding of the heuristic distinction between fixed and circulating capital. It may thus appear expedient to substitute the notion of convertibility for the potentially misleading distinction between them [Mises, 1966: 504]. In this view, fixed and circulating capital goods generally stand at two poles of the range of convertibility, where the former capital goods tend to be more convertible than the latter. Understood thus, and only thus, a differentiation between the two types of capital can still play a useful role in the theory of capital.

Capital goods may be classified according to their degree of convertibility. This classification depends crucially on plan revision. The first group of capital goods includes those that may become useless as a result of plan revision. They will be disposed of—leased out, sold, or even abandoned—because they cannot be converted to serve new production purposes. The second group includes those that may require more or less extensive conversion. Finally, the third group includes those capital goods that may be employed in the revised plan without any conversion. They are versatile in the sense that they may serve as an input in both production processes, before and after plan revision.[13]

It can be readily seen that buildings—customarily classified as fixed capital assets—are, in fact, often designed to be convertible to new uses, if not outright versatile. This holds especially for their structural systems and the basic mechanical and electrical systems, such as heating, ventilating, and air-conditioning systems, water and sewage systems, and vertical transportation systems. Most other building components, such as interior partitions, furnishing, communications systems, or electrical wiring, can be readily rearranged to fit new needs, or replaced altogether. Although they tend to be specific to a particular function, they are hardly fixed even in the short run, due to maintenance and replacement activities.

Buildings would thus generally fall in the last two categories outlined above. In fact, it may be hypothesized that the most durable capital goods or their components are as a rule designed to be most convertible, that is, adaptable and functionally flexible. Of course, in the case of buildings this holds only for a *given* location, as they are "locationally inconvertible" capital goods *par excellence*. This subject will be discussed in greater detail in Chap. 4, where we will focus on capital restructuring in the context of business and building cycles.

EX ANTE AND EX POST SUBSTITUTABILITY

As we have seen, the distinction between complementarity and substitutability of heterogeneous capital goods is associated with plans and plan revisions, respectively.[14] However, plans typically include a set of contingencies [Lewin, 1986: 212]. Some events will require minor adjustments to the plan, while others demand more drastic measures. Drastic measures are required when there are no adequate contingencies in the original plan. Therefore, for a range of possible changes, the very notion of contingency planning in fact implies the anticipation of plan revision.

Let us concentrate on durability and specificity of capital goods in the context of contingency planning. Suppose that the more durable a capital good is, the less specific its design will tend to be, at least in terms of its most durable components. A sensible designer would accommodate the unknowable future by allowing for unanticipated contingencies. The more durable a capital good is, the more a plan involving it is likely to be affected by unexpected change. Therefore, contingency planning will require a lower level of specificity for durable capital goods, including buildings.

The planner actually faces a trade-off here. The more specific a capital good, the greater the loss should his or her plan fail either partially or completely, because conversion is costly. Of course, disposition is also costly. The less specific a capital good, the greater the loss in the case of plan success, because the provision of adaptability and flexibility involves additional costs, as well. The latter possibility appears to be more critical than the former, at least at the inception of the plan, when its success is anticipated. The remaining two possibilities are obvious.[15] Clearly, the choice of the level of specificity of capital goods, and especially the most durable ones, will reflect the optimism or pessimism of the planner regarding the expected fulfillment of the plan.

This argument may be extended from plan revision to wholesale plan changes, leading to another useful distinction between two types of potential buyers of capital goods rendered useless and disposable because of plan failure.

First, the planner may decide to make the durable capital goods specific enough to fit the plausible alternative plans should the present plan fail. Because they fit the field of production the firm is currently engaged in, these capital goods may appeal to the firm's competitors also. Second, the planner may make the capital goods so nonspecific or adaptable that they are potentially attractive to others outside the firm's field of production. This would further increase their potential marketability.

Let us now explore the connection between convertibility and versatility, on the one hand, and substitutability, on the other, with specific reference to buildings. In general, convertibility and versatility concern the degree to which a given input can be placed at the service of alternative production goals, whereas substitutability refers to the degree to which the given purposes of a specified input can be equally served by some other input. Clearly, buildings can often serve as substitutes for one another. This is even more true of smaller units of useful space, such as rooms or floors, as they tend to be more homogeneous than buildings themselves. Therefore, new buildings are likely to compete with the existing ones, especially in the same geographic area. In connection with our discussion, we should bear in mind that this is, in fact, a consequence of the convertibility and versatility of buildings.

Because such issues affect contingency planning, it is clear that a rigid distinction between complementarity and substitutability is unwarranted. While complementarity of heterogeneous capital goods may be the central issue in a plan, and substitutability the central issue of plan revision, both notions are likely to be found in the original plan itself. The possibility of plan failure exists from the very inception of a plan. Coping with this possibility includes gauging the *potential* substitutability of capital goods, that is, the degree to which they may compete with each other under different economic conditions. This is especially important in the case of buildings, the most durable among capital goods. For example, plan failure may require moving a part of the production equipment to a smaller building, which should be adapted to the new function well in advance of the actual move. The same holds for the vacated building with respect to another new function. As the two buildings are (imperfect) substitutes for one another, contingency planning should concern a range of their alternative uses in anticipation of plan failure. As we have seen in Chap. 1, the above argument should be understood dynamically; as part of a continual adjustment to changing economic conditions, not as a problem confined to the planning and design stage of the building process.

It therefore appears useful to distinguish between *ex ante* and *ex post* substitutability of capital goods. *Ex ante* substitutability becomes an aspect of contingency planning and reflects anticipated change. *Ex post* substitutability is an aspect of plan revision *sensu stricto* and concerns unanticipated change. The more durable a capital good, the more important *ex ante* substitutability is likely to be in the mind of the planner. This is especially important when rapid economic change may be anticipated with certainty, but not in sufficient detail.

At any rate, the distinction between complementarity and substitutability of capital goods cannot stand as initially proposed.

We will return to this topic in Chap. 3, where we will apply the concepts of complementarity and substitutability to different parts of the same building, that is, building components. There, complementarity and *ex ante* substitutability will be associated with the building design, whereas *ex post* substitutability will be linked to building management, especially as it relates to building maintenance and building component replacement.

OPERATING ASSETS AND SECURITIES

All capital goods are heterogeneous. We therefore encounter the problem of measuring capital, broached in preceding sections. In search of a common denominator, economists almost inevitably follow accountants in adopting money value as the standard of measurement of capital goods. Capital can be measured in this fashion only in stationary equilibrium, which is itself only a scaffolding for analysis, a useful fiction, as it were. However, the common denominator is lost whenever relative money values change [Lachmann, 1976b: 152–53, and 1978a: 2, following Hayek, 1941]. In disequilibrium, where money values are not consistent with each other, capital cannot be adequately measured. A generic concept is nevertheless needed to classify capital resources even if we cannot measure them.[16]

This classification or structural order must rest on the complementarity of the means employed in a given area of action for distinct ends. The concept of human action must inform the classification.[17] In this context, we are also concerned with the existence of capital complementarity outside the sphere of physical capital goods. Put differently, there should be complementarity in a well-selected investment portfolio [Lachmann, 1978a: 86].

The classification of capital assets can be presented as a tree, shown in Fig. 2.1.[18] Some clarification is needed here. Operating assets are the physical capital goods and money complementary to them, while securities are the titles which embody the control of production, as well as define the recipients of payment. First-line assets are those capital goods whose services provide the input of the production plan right from the start, whereas second-line assets are those that, like spare parts, or money for wage payments, are intended to be put into operation at a definite point in time during the plan period. Reserve assets are those, like the cash reserve and reserve stocks, that will not have to be used at all if the plan succeeds. They will therefore tend to be more liquid, that is, more readily exchangeable, than other assets committed under the production plan. Finally, debt-titles are securities embodying the right to an income in terms of currency units, and equities embody the right to participate in control and in residual income.[19]

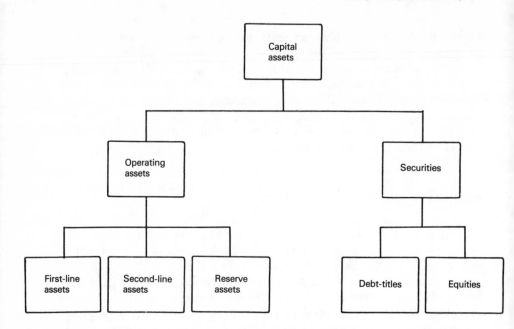

Figure 2.1 Classification of capital assets.

It should be noted that buildings (and parcels of land they occupy) generally fall into the category of first-line assets. Some buildings represent reserve assets, however. Although entire buildings may be deliberately designated as
reserve assets in a production plan, this is much more likely to be a result of
malinvestment in the past. The concept of reserve assets is more applicable to
space, rather than to buildings as such. As we will see in Chap. 5, a significant
proportion of space at the disposal of an economic agent may be in this category
at any one time.

This classification offers a basis for the general classification of capital
structure. Three kinds of structure can be distinguished: the plan structure concerning technical complementarity, the control structure based on the relationship between debt and equity, and the portfolio structure depending on investors'
asset preferences [Lachmann, 1978a: 91]. Whether a production plan is feasible
depends also on the investors' willingness to acquire the securities necessary to
finance the plan. This will, in turn, depend on the relationship between debt and
equity offered to the investors. In this scheme, the manager's production plan
concerns operating assets, while the capital owners' plans concern securities
[Lachmann, 1978a: 98].[20]

The analysis can now proceed to the dynamic forces that bring the plan
structure, the control structure, and the portfolio structure into consistency with
each other. The key role in this process is assigned to the stock exchange, often
called "the market" because of its central role in market economies. The stock

exchange is a market not for capital goods but for titles to them. It is an instrument for promoting consistent capital change [Lachmann, 1978a: 86].

PLANS AND SECURITY PORTFOLIOS

The plan structure remains undisturbed as long as everything goes "according to plan." That is, as long as reserve assets neither increase nor decrease significantly; operating cash balances and stocks are continually replenished out of gross revenue; and a steady yield stream in the form of money payments flows from cash balances to the holders of securities. In this case investors will have no incentive to change the composition of their security portfolios. Under these conditions we have a stationary equilibrium [Lachmann, 1978a: 92].

However, it is unlikely that everything will go according to plan. Both unexpected success and failure may upset the balance of operating assets leading to a change in their composition. Expansion following success will very likely require new investment.[21] Of course, the plan may also fail. Temporary failure may involve a short-term drain of reserves without affecting the balance of capital assets. The greater the liquidity of the reserve assets, the better. "Liquidity is freedom," as Hicks [1979: 94] states emphatically. If the failure is severe, however, it may become impossible to maintain or to replace capital assets as they wear out. In such a case, a plan revision and a reshuffle of capital assets may become necessary. This will almost certainly involve new investment. Thus, both expansion following success and reconstruction following failure lead to an increase in the "demand for money" [Lachmann, 1978a: 93]. This argument will be elaborated in Chap. 4 in the context of business cycles.

Now, capital gains and losses accompany the success and failure of production plans, respectively. The stock exchange comes into play at this juncture. Capital gains and losses modify the portfolio structure, that is, the relative values of portfolio components [Lachmann, 1978a: 94–95]. Market processes involving transmission of knowledge bring the various components of the asset structure into consistency with each other. They also modify the control structure and the composition of investment portfolios. Evaluation of securities by the market plays a key role in these processes. In fact, it can be said that a stock exchange is indispensable in a market economy [Lachmann, 1977: 161]. It consists of a series of markets for assets, that is, of future yield streams. In each market supply and demand are brought into equality every market day.

This process depends crucially on the divergence of expectations of different market participants. Unless expectations concerning the future yield stream of a productive asset differ there can be no market at all, because transactions take place only between those whose expectations diverge from the current market price [Lachmann, 1977: 161, and 1986: 18]. As all assets traded on the stock exchange are (imperfect) substitutes for one another, these markets form a system. Equilibrium is attained simultaneously in each market. In this way the market economy accomplishes a consistent evaluation of all its major productive assets. Because investors' expectations are continuously revised as new knowledge is acquired, the market equilibrium has to be reestablished each market day.[22]

The continual revaluation of productive assets, as reflected in portfolios of securities, reflects the optimism or pessimism of portfolio holders concerning production plans undertaken by specific firms. Individual investment projects, let alone individual capital assets forming a capital combination, generally cannot be disentangled from each other. As Lachmann [1976b: 155] observes, "[t]he objects of valuation are not individual capital goods but fractions of capital combinations that are the substrate of multiperiod plans." Similarly, it is realistically impossible to disentangle profits from capital gains. The plan itself, reflected in the plan structure, is the basic unit of analysis.

THE VALUE OF COMPLEMENTARY CAPITAL GOODS

Consider a building that is a part of a capital combination supporting a particular production plan. Such a capital combination may include several buildings. All capital goods included in the capital combination are complementary in the sense that they all contribute to the production process in accordance with a plan. One question that may arise in this context is: What is the value of a particular building forming a part of a capital combination? This is the problem of imputation, which is the basis for determining prices of capital goods, as well as the distribution of yield or income these goods jointly help generate in the process of production.

It should be noted in passing that this question does not contradict the previous claim that capital goods defy aggregation, that is, objective measurement by an outside observer. Individuals of course make attempts to gauge the value of particular capital goods, especially when these goods are being sold or bought. As Kirzner [1976b: 140] observes, "[w]hat cannot be 'objectively' measured by the outsider turns out to be evaluated subjectively by the relevant decision maker." This distinction will be especially useful in Chap. 5, where we will return to the problem of measurement in the context of real property management tools and approaches.

In general, the value of higher order goods is determined by the prospective value of the lower order goods they help produce [Menger, 1981: 150].[23] The value of a particular higher order good is equal to the prospective value of the portion of the quantity of a good of lower order that would not be produced if the higher order good were unavailable in the production process [Menger, 1981: 164].[24] In modern terminology, the value of a particular capital good is equal to the prospective value of its marginal product. Note the emphasis placed on prospective value. By implication, the value of capital goods is always in an essential sense speculative, as it depends on expectations concerning future states of the economy. As Keynes [1964: 149–50] argues,

> The outstanding fact is the extreme precariousness of the basis of knowledge on which our estimates of prospective yield have to be made. Our knowledge of the factors which will govern the yield of an investment some years hence is usually very slight and often negligible. If we speak frankly, we have to admit that our basis of knowledge for estimating the yield ten years hence of a railway, a copper mine, a textile factory, the goodwill of a patent medicine, an Atlantic liner, a building in the City of London amounts to little and sometimes to nothing; or even five years hence. In fact, those who seriously attempt to make any such estimate are often so much in the minority that their behavior does not govern the market.

The value of a building, as a complementary capital good forming part of a capital combination, will be determined by the loss of return that might be suffered in its absence. However, the value of goods the building helps produce is at any moment only a prospective value, that is, the value of goods that will be sold at some future date. It is determined on the basis of someone's judgment about future market conditions. Alternatively, it is determined by the judgment of which different wants a particular product may satisfy. Note that buildings in particular are by and large designed so that they may serve several alternative productive purposes, which further complicates the problem of imputation.[25]

Now, in the capital budgeting literature it is customary to define the value of a complementary good by its replacement cost. This does not contradict the above argument. It merely refers to a special condition, which does not always pertain. Valuation according to replacement cost is not permissible where prompt replacement through reproduction is not assured or in the case of irreplaceable goods [Böhm-Bawerk, 1959c: 103]. Thus we use two different valuations of reproducible goods—by the value of their marginal product and by their replacement cost—each of which valuations is made conditionally, that is, each of which pertains under different assumptions.[26]

In the case of buildings, these alternative assumptions about replacement lead to further complexities. A building may be replaced in several ways. An existing building may be purchased or leased and used in its present condition. Furthermore, an existing building may be purchased and renovated, in which case the process of renovation would take time. Finally, a new building may be ordered, which would again require time for construction. In the latter two cases, the replacement cost of an existing building would have to include the value of the product lost during the period of renovation or construction, respectively. Of course, all of these options would have to be considered as alternatives on the basis of which the replacement cost might be established and the valuation of a particular building made.

CAPITAL, MONETARY CALCULATION, AND BOOKKEEPING

Monetary calculation guides entrepreneurial action in the market process. By calculating, the entrepreneur distinguishes profitable from unprofitable lines of production. In other words, those of which consumers are likely to approve from those they are likely to disapprove. Looking forward, the entrepreneur is engaged in commercial precalculation of expected proceeds and costs. Looking backward, the entrepreneur establishes the outcome of past action through accounting of profit and loss.

It is worth noting in passing that calculation is always essentially speculative. As Keynes [1964: 150] writes, "[i]f human nature felt no temptation to take a chance, no satisfaction (profit apart) in constructing a factory, a railway, a mine or a farm, there might not be much investment merely as a result of cold calculation." Keynes [1964: 161–63] calls this spontaneous urge to action "animal spirits." However, Keynes [1964: 162–63] adds:

> We should not conclude from this that everything depends on waves of irrational psychology. [. . .] We are merely reminding ourselves that human decisions affecting the future, whether personal or political or economic, cannot depend on strict mathematical expectation, since the basis for making such calculations does not exist; and that it is our innate urge to activity which makes the wheels go round, our rational selves choosing between the alternatives as best we are able, calculating where we can, but often falling back for our motive on whim or sentiment or chance.

Monetary calculation permeates the market economy. It is the main vehicle for planning and acting in a society directed and controlled by the market and its prices. Monetary calculation reaches its zenith in capital accounting [Mises, 1966: 230]. Bookkeeping is central to calculation and to the market process. It may be considered a great achievement of human development.[27]

Like monetary calculation discussed above, bookkeeping necessarily contains a subjective element concerning future market conditions. As Mises [1966: 213-14] writes:

> The dully kept accounts in a system of correct bookkeeping are accurate as to dollars and cents. They display an impressive precision, and the commercial exactitude of their items seems to remove all doubt. In fact, the most important figures they contain are speculative anticipations of future market constellations.

Concerning the illusory precision of bookkeeping and the fact that the valuations of goods are always based on estimates depending on uncertain or unknown factors, Mises [1980: 234] elaborates:

> So far as this uncertainty arises from the commodity side of the valuations, commercial practice, sanctioned by the law, attempts to get over the difficulty by the

exercise of the greatest possible caution. With this purpose it demands conservative estimates of assets and liberal estimates of liabilities [. . .].

This is of particular interest for long-lived capital goods, such as buildings,
where the overestimation of future costs and the underestimation of future
proceeds is likely to be especially pronounced due to the commercial practice
outlined above.

The subjective aspect of bookkeeping is of great significance in general.
In particular, the value of capital goods depends on the expected costs and
proceeds of production, which ultimately depend on the expected market conditions, that is, on consumers' valuations of the future products. The rules of
bookkeeping may constrain a full recognition of the anticipated events. This implies that the book value and market value of durable capital goods will in
general be different. The stock exchange will therefore tend to disregard the
book value of capital goods.[28] It goes without saying that the book value may be
of considerable interest in connection with both income and real estate taxes
levied from business enterprises.[29]

Of course, stock exchange appraisals of the market value of capital goods,
that is, aggregates of such goods owned by particular enterprises, fluctuate over
time. At times the reappraisals are rather violent, precipitating speculative take-
overs capitalizing on the discrepancy between the book value and market value
of capital goods owned by a particular enterprise. At any rate, the stock exchange helps bring the book value and market value of capital goods in closer
conformance with each other.

Monetary calculation and bookkeeping make it possible to relieve the
entrepreneur of involvement in excessive detail. Profit and loss in particular
production lines becomes visible at a glance. In a sense, each division of an
enterprise "owns" a definite part of its capital, and thus it is possible to calculate
profit and loss associated with each division.[30]

In such a system, the entrepreneur can assign a great deal of autonomy to
each division's management. An examination of the accounts shows how successful the management of each division was. The managerial function is therefore subservient to the entrepreneurial function. The primary function of the
entrepreneur is to direct the employment of factors of production for the ac-
complishment of definite tasks. Controlling these factors ultimately brings the
entrepreneur either profits or losses, created through the discrepancy between the
expected prices and the prices actually established in the market [Mises, 1966:
294].[31]

In a growing economy, in which the per capita share of capital invested is
increasing, the range of entrepreneurial activities includes the determination of
the employment of the additional capital goods accumulated by new savings.
The entrepreneur determines the financial structure by determining in what lines
of business to employ new capital, as well as how much new capital to employ
[Mises, 1966: 307]. The accounting system will be instrumental in the allocation

of the additional capital goods, as well. Profitable lines of production will tend to get a larger share than less profitable lines. Clearly, the additional capital goods may also be used in combination with the existing capital goods employed in the unprofitable lines of production in order to promote a new and potentially profitable line of production.

CONCLUSION

There are several key propositions presented in this chapter. They were discussed in Chap. 1, but only in a preliminary way. First, buildings and building processes cannot be understood outside the underlying economic process as a whole. Most important, this pertains to the value of real property, buildings and land. This is crucial in view of the tendency to reduce the problem of building economy to that of economizing the use of scarce resources in the building process *sensu stricto,* that is, in the construction process.

Second, the value of real property ultimately depends on the value of goods and services produced with its help. However, this value is only prospective, subject to radical revaluation. By the same token, it is subjective in nature. There is no reason to believe that two individuals will readily agree in their evaluations of the prospective value of a production process, including the buildings and land involved.

Third, production plans may fail, and economic subjects are fully aware that they may indeed fail. For that reason they will frame their production plans in such a way that capital combinations at their disposal, including real property, will withstand changes in economic conditions. The flexibility and adaptability of buildings and other capital goods is one of the most important characteristics of a resilient production plan.

With this background, we are ready for the discussion of the building process itself. In Chap. 3 we will focus on building costs, but we will show that value and cost concepts have much in common, as they are linked by the concept of opportunity cost. Although the next chapter will also contain several sections that some readers may find overly abstract, it will address building issues more directly than this chapter did.

APPENDIX: "ANGLO-AMERICAN" (A) AND "AUSTRIAN" (B) PROPOSITIONS ON CAPITAL THEORY, ACCORDING TO HAYEK [1941: 47–49]

1A. Stress is laid exclusively on the role of fixed capital, as if capital consisted only of very durable goods.

1B. Stress is laid on the role of circulating capital that arises out of the duration of the production process, because this brings

2A. The term capital goods is reserved for durable goods that are treated as needing replacement only discontinuously or periodically.

3A. The supply of capital goods is assumed to be given for the comparatively short run.

4A. The individual durability of a capital good is assumed to be the relevant time factor that we need to consider to understand the effect of changes in the rate of interest on the value of that capital good.

5A. The production technique employed is assumed to be determined by the given state of technological knowledge.

6A. The need for more capital is assumed to arise mainly out of a *lateral* expansion of production, that is, a mere duplication of equipment of the kind already in existence.

7A. The change that will initiate additions to the stock of capital is sought in an increase in *absolute* demand, that is, in the total money expenditures on consumers' goods.

8A. In order to explain a lateral expansion of production, the existence of unemployed resources of all kinds is postulated.

out particularly clearly some of the characteristics of all capital.

2B. Nonpermanence is regarded as the characteristic attribute of all capital goods, and the emphasis is accordingly laid on the need for continuous reproduction of all capital.

3B. It is assumed that the stock of capital goods is being constantly used up and reproduced.

4B. It is not the durability of a particular good that is regarded as the decisive factor, but the time that will elapse before the final services to which it contributes will mature. That is, it is not the attributes of the individual good but its position in the entire time structure of production that is regarded as relevant.

5B. The availability of capital at any time is assumed to determine which of many production techniques will be employed.

6B. Additional capital is assumed to be needed for making changes in the production technique (that is, in the way in which individual resources are used), and to lead to *longitudinal* changes in the structure of production.

7B. Changes in the stock of capital are assumed to be determined by changes in the *relative* demand for consumers' and producers' goods, respectively.

8B. In order to stress the changes in production technique connected with an increase of capital, the existence of full employment is usually postulated.

9A. The demand for capital goods is assumed to vary in the same direction as the demand for consumers' goods but to a greater degree.

9B. The demand for capital goods is assumed to vary in the direction opposite to the demand for consumers' goods.

10A. The analysis is carried out in monetary terms, and a change in demand is assumed to mean a corresponding change in the size of the total money stream.

10B. The analysis is carried out in "real" terms, so that an increase in demand somewhere must necessarily mean a corresponding decrease in demand somewhere else.

NOTES

1. The classification of goods into first order goods (or consumption goods) and higher order goods (or production goods) was introduced by Menger [1981]. This classification is useful for an understanding of economic value. For a comprehensive review of the historical and philosophical roots of Menger's theory of value, see Grassl and Smith [1986], for example.

2. This is one of the central ideas introduced by Menger [1981]. More generally, as Kirzner [1979: 162] observes, "Menger's Law—as we may call this insight—draws attention to man's propensity to attach the value of ends to the means needed for their achievement." It follows that value is subjective in essence. It is inseparable from human ends. As Menger [1981: 120–21] points out:

   Value is thus nothing inherent in goods, no property of them, nor an independent thing existing by itself. It is a judgement economizing men make about the importance of the goods at their disposal for the maintenance of their lives and well-being.

It should be noted, however, that Menger seems to have regarded the nature of human needs as objective, that is, independent of human choice. For example, Menger [1985: 63] writes:

   The most original factors of human economy are the needs, the goods offered directly to humans by nature [. . .], and the desire for the most complete satisfaction of needs possible [. . .]. All these factors are ultimately given by the particular situation, independent of human choice.

According to Lachmann [1978c: 57], Menger's subjectivism is thus incomplete, because value is subjective not only with respect to the means, but also the ends.

3. The Appendix is based on Hayek [1941: 47-9]. It offers several insights useful throughout this chapter. In addition to the contrasting views on capital durability, most important are the contrasts between the role of fixed and circulating capital, the relevant time factor, and the role of maintenance and replacement in the reproduction of capital.

4. This overview is based on Kirzner [1966: Chap. 1].

5. As Kirzner [1966: 4] writes:

> The market process consists in the systematic chain of events that ensue from the interaction in the market of the decisions of numerous individuals. These decision-makers find, at the outset, that the opportunities that are in fact being offered to them in the market [. . .] are different (either more or less attractive) than those originally expected. From the lessons learned in the market in this way, there follows a systematic pattern of adjustment and revision in the decisions made by market participants. This pattern of adjustment constitutes the market process.

6. According to Kirzner [1966: 31]:

> [I]t is precisely this, the enforcement of changes in plans in order to bring different plans into closer mutual adjustment, that is the function of the market process. This is well recognized in the case of single period plans. What is here being pointed out is that in the context of multi-period planning, the market process exercises a very special influence. This influence consists in the enforcement of new plans that will, possibly, embrace uses for existing tangible things for which these things had originally not been constructed.

7. As Kirzner [1966: 41] argues:

> In this view of the role of capital theory, we join those writers who have deplored the long standing, almost traditional, tendency to see capital theory as a mere adjunct to the theory of interest. Interest rates represent merely one particular kind of price—the intertemporal price—that is determined, along with all the other market phenomena, by the multi-period plans that are being made by market participants. Of course capital theory is involved. But capital theory is required in order to understand the current market price for shoes, just as much as it is required in order to understand the ratio between the future price of shoes and the current prices.

Here Kirzner follows the lead of Hayek [1941] and Lachmann [1978a], who have insisted on the distinction between the theory of capital as such and its subservient uses in the context of the theory of interest. (See note 2 in Chap. 1.)

8. This overview is based on Lachmann's [1978a] theory of capital, and on Lewin's [1986] presentation of Lachmann's theory. As this section deals with standard Austrian concepts, no further references will be given. Useful overviews of the Austrian capital theory can be found in Lachmann [1976a] and Kirzner [1976b and 1976c], for example.

9. According to Lachmann [1978a: 6–7]:

> Once we allow for heterogeneity we must also allow for complementarity between
> old and new capital. The effect of investment on the profitability of old capital is
> now seen to depend on which of the forms of old capital are complementary to, or
> substitutes for, the new capital. The effect on the complements will be favorable,
> on the substitutes unfavorable.

10. Capital maintenance and replacement have been typically treated as routine activities,
    whereas the theory of capital presented here considers them central for capital forma-
    tion in a changing world. According to Lachmann [1978a: x]:

> We now realize that capital replacement, far from being a matter of business
> routine, is a most problematic activity. It must rest on expectations, subjective and
> individual, about future income streams and choice among them. There can be no
> such thing as a "correct" method of depreciation and replacement in a changing
> world.

Of course, the pattern of interlocking capital maintenance and replacement plans is of
special concern to us here. (See note 12 in Chap. 1.)

11. It is interesting to note that this argument can be traced to Ricardo [1951: 31n, 150],
    who pointed out that there is no definite line of demarcation between circulating and
    fixed capital. Consequently, this division is not essential [Ricardo, 1951: 31n].

12. As Mises [1966: 504] argues:

> The more a definite process of production approaches its ultimate end [that is,
> goods produced for consumption], the closer becomes the tie between its inter-
> mediary products and the goal aimed at. Iron is less specific in character than iron
> tubes, and iron tubes less so than iron machine parts. The conversion of a process
> of production becomes as a rule the more difficult, the farther it has been pursued
> and the nearer it has come to its termination, the turning out of consumers' goods.

13. Mises [1966: 503] distinguishes between four groups of capital goods according to
    their degree of convertibility, as follows:

> If in the course of the period of production the goal is changed, it is not always pos-
> sible to use the intermediary products already available for the pursuit of the new
> goal. Some of the capital goods may become absolutely useless, and all expenditure
> made in their production appears now as waste. Other capital goods could be util-
> ized for the new project only after having been subjected to a process of adjust-
> ment; it would have been possible to spare the costs required by this alteration if
> one had from the start aimed at the new goal. A third group of capital goods can be
> employed for the new process without any alteration; but if it had been known at
> the time they were produced that they would be used in the new way, it would have
> been possible to manufacture at smaller cost other goods which could render the
> same service. Finally there are also capital goods which can be employed for the
> new project just as well as for the original one.

14. Following Lachmann [1978a], Lewin [1986: 213] writes:

> Whatever its motivating cause, a plan revision entails the *substitution* of some resources for others. Substitutability is a phenomenon of change. [. . .] Thus, while complementarity is an aspect of any given plan, substitutability is an aspect of contemplated *changes* to the plan. Together these two concepts characterize different aspects of the capital structure, namely its coherence and its adaptability.

15. This trade-off is reminiscent of Type I and Type II errors in statistical hypothesis testing, where the erroneous rejection of the correct hypothesis is often treated as less desirable than the erroneous acceptance of the incorrect hypothesis.

16. As Lachmann [1978a: 4] writes:

> We want to be able to speak of "Capital." Logically, we can establish no systematic generalization without a generic concept. Unable as we are to measure capital resources, we must at least make an attempt to classify them. If there can be no common denominator there should at least be a *criterion ordinis*.

17. According to Lachmann [1978a: 89]:

> Since [. . .] our purpose is praxeological, not merely taxonomic, since our interest is in assets *qua* instruments of action and the structural relationships between them as channels for the transmission of knowledge, our mode of classification is governed by the relevance of our classes to planning and action.

> Praxeology is concerned with the purposefulness and rationality of human action. For a systematic discussion of the praxeological conception of economic science, see Mises [1966 and 1978]. Kirzner [1976a: Chap. 7] provides a useful overview of the historical development of praxeology.

18. For further detail, see Lachmann [1978a: 89-91].

19. It should be emphasized that the distinction between fixed and circulating or working capital does not figure in this classification of operating assets. By implication, such a distinction is of little praxeological value in this connection. (See note 11.)

20. It should be noted that this presentation assumes a simple type of asset structure in which all securities directly "represent" operating assets. This need not be so; securities may exist that "represent" other securities [Lachmann, 1978a: 96]. Similarly, it is assumed here for simplicity that all operating assets are directly owned, rather than rented or leased. This need not be so, either. Concerning the real property, for example, some buildings and parcels of land may be leased from others and used as parts of a capital combination. These complexities will not detain us here; however, we will discuss the use of leased space in Chap. 5.

21. As Lachmann [1978a: 92-3] writes:

> If success is unexpectedly great problems begin to arise. The surplus profits [. . .] have to be assigned to somebody. They may be used for higher dividends or be "ploughed back" or serve to pay off debt. In the first case they will, in addition to

giving higher incomes, entail capital gains to shareholders, and hence change the total value as well as composition of their portfolios. In the second case they will induce and make possible a new plan structure. In the third case they will modify the control structure.

In this context it is interesting to briefly explore the relationship between investment risk and the control structure. As Lachmann [1978a: 93n] points out:

A degree of risk does not attach to a given investment project as such, but always depends on the control structure. There are many projects which a young and heavily indebted firm would not dare to touch, but which an old firm with low debt and ample reserves can afford to take in its stride.

This is important in view of the common practice in so-called risk analysis. There, an evaluation of risk is typically assigned to investment projects as such.

22. As Lachmann [1977: 162] argues:

For precisely the same reason for which equilibrium in an asset market is reached so smoothly and speedily, it cannot last longer than one day. For expectations rest on imperfect knowledge, and not even a day can pass without a change in the mode of diffusion of knowledge.

23. As was already mentioned, goods of first order are those that satisfy our immediate needs, while goods of higher orders participate in the provision of means for the satisfaction of our direct needs at various stages of production. (See notes 1 and 2.)

24. Böhm-Bawerk [1959c: 80] concurs with Menger's approach: "Menger and I [...] ascribe to each complementary good as the basis for its valuation the full loss of return which its absence would cause." Böhm-Bawerk [1959c: 85] elaborates:

What must be foregone in case of the loss of a good is always and necessarily identical with that which is attained by its possession. They are two different forms of perception for one and the same thing. [. . .] [W]e value goods in our possession according to the loss we would suffer in case of their absence, and we value goods which we should like to acquire according to the increment of utility which their acquisition and possession will render us.

25. According to Böhm-Bawerk [1959c: 97]:

In discussing the price determination of producers' goods suitable for many employments I presumed it is possible or even typical that various goods producible with the same kind and quantity of producers' goods can obtain a *different* marginal utility and value by serving the gratification of different wants. On this basis I advanced the formula that the value of a producers' good is determined by the marginal utility and value in its *least valuable* product.

Hayek [1984: 37] elaborates: "[a]ccording to this principle, the value of a good employed in a certain use is not always equal to the minimum level of utility realized in this use, but to the minimum level of utility which may be realized in any of the uses which may be made of it."

26. As Böhm-Bawerk [1959c: 104] writes:

We value according to costs if replacement through reproduction can be considered as assured. And we value according to the higher direct marginal utility if the replacement cannot or cannot yet be considered as assured. We need both kinds of valuation in order to be correctly informed on each situation which may induce us to act.

27. Mises [1966: 230] observes that "Goethe was right in calling bookkeeping by double entry 'one of the finest inventions of human mind'."

28. According to Mises [1966: 349]:

Commercial usages and customs and commercial laws have established definite rules for accountancy and auditing. There is accuracy in the keeping of the books. But they are accurate only with regard to these rules. The book values do not reflect precisely the real state of affairs. The market value of an aggregate of durable producers' goods may differ from the nominal figures the books show. The proof is that the Stock Exchange appraises them without any regard to these figures.

29. For a discussion of tax issues in the U.S. regarding capital goods in general, see, for example, Feldstein [1987]. For a discussion of the differential effect of tax rules on structures and equipment, as well as on industrial and commercial structures, see Hines [1987].

30. As Mises [1966: 305] argues:

It is the system of double-entry bookkeeping that makes the functioning of the managerial system possible. Thanks to it, the entrepreneur is in position to separate the calculation of each part of his total enterprise in such a way that he can determine the role it plays within the whole enterprise. Thus he can look at each section as if it were a separate entity and can appraise it according to the share it contributes to the success of the total enterprise. Within this system of business calculation each section of a firm represents an integral entity, a hypothetical independent business, as it were. It is assumed that this section "owns" a definite part of the whole capital employed in the enterprise, that it buys from other sections and sells to them, that it has its own expenses and its own revenues, that its dealings result either in profit or in loss which is imputed to its own conduct of affairs as distinguished from the results of the other sections.

31. But Mises [1980: 234–35] warns that one shortcoming of accounting is that it usually does not take into account the variability of the value of money itself. In business practice, money is customarily held to be a straightforward measure of price and value. This is especially troublesome in the accounts of profit and loss, which may have a direct impact on capital consumption. As Mises [1980: 235–36] argues:

This disregard of variations in the value of money in economic calculation falsifies accounts of profit and loss. If the value of money falls, ordinary bookkeeping, which does not take account of monetary depreciation, shows apparent profits, because it balances against the sums of money received for sales a cost of production calculated in money of a higher value, and because it writes off from book values originally estimated in money of a higher value items of money of a smaller value. What is thus temporarily regarded as profit, instead as part of capital, is consumed by the entrepreneur or passed on either to the consumer in the form of price reductions that would not otherwise have been made or to the laborer in the form of higher wages, and the government proceeds to tax it as income or profits. In any case, consumption of capital results from the fact that monetary depreciation falsifies capital accounting.

Therefore, a fall in the purchasing power of money may lead to an unwitting increase in capital consumption or depreciation, and vice versa. In both cases, ordinary bookkeeping falsifies accounts of profit and loss by abstracting from the variation in the purchasing power of money; bookkeeping thereby deforms the capital structure.

3

Building Process

INTRODUCTION

In the presentation of the building process we will concentrate mainly on various components and aspects of total building cost. This is the vantage point customarily taken in the building economics literature. However, we will also consider building value. The foundation needed to understand both value and cost was provided in Chap. 2.

Economic change is pervasive. Our main task in this chapter is to examine the effect of economic change on building utilization and operation. However, all phases of the building process—beginning with planning, design, and construction—are susceptible to unexpected shifts in economic conditions experienced by the building owner. One of the key problems of the building realm therefore concerns the flexibility and adaptability of buildings—their fabric, configuration, and structure—during the entire building process. The extent to which buildings can be modified to meet the needs of their owners ultimately depends on the state of building technology, as we will argue in Chap. 4. There, we will also be concerned with economic change, but we will focus on systematic changes that affect an economy as a whole.

A broad array of subjects will be analyzed in this chapter. First, we will discuss the value of the three complementary factors of production—land, labor, and capital. We will emphasize that their relative values vary over time, depending on the particular circumstances of time and place. Second, we will address the nature of economic knowledge from the standpoint of an economizing individual. Considerations of time and place are especially relevant to real property, as it involves an enormous amount of detailed and fragmented knowledge accessible to a large number of widely dispersed and loosely interrelated individuals involved in the building process. Third, the problem of real property and the role of government intervention will be briefly discussed in the context of external costs and benefits of real property undertakings. Fourth, we will argue that a subjectivist conception of costs is indispensable in the building realm, primarily because it provides the link between cost and value concepts. The concept of opportunity cost will play an essential role in several sections dealing with the building process. We will focus on the building owner in the context of ever-changing economic conditions. Fifth, some issues of construction productivity and quality of construction goods will be analyzed, with an emphasis on their flexibility and adaptability. Sixth, we will return to the relationship between fixed and circulating capital from the vantage point of buildings as capital goods. Maintenance and replacement issues will be treated in this context, as well. Seventh, the continually changing relationship between the use value and exchange value of buildings will be discussed. Eighth, the interaction between economic fluctuations and the building process will be introduced, in preparation for Chap. 4. The emphasis will be placed on maintenance and replacement activities in the context of building utilization and operation.

LAND, LABOR, AND CAPITAL

Natural forces continually transform matter. Some useful forms of matter come into existence fortuitously and without human intervention; however, the vast majority of things require conscious and purposive intervention into the natural process. This activity consists of guiding the natural process in accordance with the rules imposed by nature. Viewed as a physical phenomenon, human productive activity is only a conversion of matter into more advantageous forms.

The ability to move matter is the key to harnessing natural forces for human purposes.[1] We intervene in the natural process solely *via* spatial control of matter. An example of this is the process by which we erect a wide variety of constructed facilities—such as roads, railways, bridges, and buildings—by transporting and assembling all the requisite building materials. Temporal control of matter is only an aspect of spatial control. Although the forces of nature are constantly in action, we can combine them so that a particular effect occurs only when desired.[2]

We have some physical power to put the forces of nature to work where and when we will, but the inert resistance of matter is often enormous. The monumental effort involved in the extraction and transportation of bulky building materials is a good example of what is required to overcome this resistance. We also experience it whenever we attempt to change or replace existing constructed facilities. This is especially evident when we are performing these tasks unaided by various types of transportation and construction equipment such as trucks, cranes, and bulldozers.

Our capacity for production depends on our intellect. The human mind has the ability to discover the causal relationships in nature, and thus it can guide natural processes toward the production of desired goods [Böhm-Bawerk, 1959b: 9]. In this way, natural processes can be made to work for us with minimal exertion of our physical power. Because nearly all powers of nature are linked with the soil, we can designate this economic natural endowment by the term *land,* or *uses of land*.[3] Land and labor are the elemental powers or factors of production. With land and labor at our disposal, we can directly satisfy some of our needs or wants.

Some useful things can be simply gathered and consumed directly; however, other consumption goods must be produced by means of production goods, which must first be acquired or produced. These production goods can be designated by the term *capital.* Capital is often designated as the third factor of production.[4] It brings two consequences, one advantageous and another disadvantageous. The indirect mode of producing consumption goods required for the satisfaction of our immediate needs tends to be more productive, but it also entails a sacrifice of time. Clearly, the production of capital goods increases the period of time, or the number of production stages, separating the production of goods and the satisfaction of human needs.

Capital goods are the focal point of the activity of the human mind. Our ability to envision the future is instrumental in this context. The forces of nature are consciously and purposefully set to work for the satisfaction of our future needs *via* capital goods. However, capital—as a factor of production—is inconceivable without land and labor. Land, labor, and capital are thus usually designated as complementary factors of production.

How should we determine the value of the three complementary factors of production? It can be assessed only in terms of their relative contributions to the satisfaction of our needs. By the same token, the theory of the value of complementary goods reveals the basis for remunerating each of the three factors of production.[5] As was argued in Chap. 2, the value of a complementary good is determined by the value of the product that would be lost if a unit of this good were removed from the process of production, that is, by the value of its marginal product.

The relative values of land, labor, and capital change over time. They differ from place to place. This is especially pronounced in the building realm. The particular circumstances of time and place provide the fundamental ingredients of the building process. Our decisions to build, to continue building, or to abandon a building already started depend on constantly changing valuations of the three complementary factors of production in their alternative uses. These alternatives can be understood only in the context of a wider economic process of which the building in question is an essential part, as has been argued in preceding chapters.

THE PARTICULAR CIRCUMSTANCES OF TIME AND PLACE

Much economic thought is preoccupied with given means and ends. If we possess complete knowledge about a given system of means and ends in the economy as a whole, the remaining economic problem is purely one of logic. The best use of the available means, including the three factors of production, can be determined by economic calculation, that is, by the pure logic of choice. However, this is decidedly not the economic problem we face in reality, because the "data" upon which the economic calculus is based are never "given" to a single mind [Hayek, 1948: 77]. Separate individuals possess dispersed bits of incomplete and often contradictory information. The knowledge of means and ends is "given" only to these individuals. The economic problem is therefore how to ensure the best use of resources known to these individuals for the satisfaction of ends whose relative importance only they know [Hayek, 1948: 78]. What applies to the society in its entirety by and large also applies to large social organizations.

The allocation of available resources, and thus all economic activity, involves planning. Even more emphatically, "all economic theory is a theory of planning" [Schumpeter, 1954: 908]. The crucial issue is who should do the planning, a central authority or many separate individuals. The most efficient economic system will be the one likely to use the existing knowledge most fully [Hayek, 1948: 79]. This issue pertains both to economy-wide central planning and "central" planning in a large private or public organization.

Centralization of knowledge tends to diminish the importance of the particular circumstances of time and place. We are often reminded that scientific knowledge relies on general rules rather than on the specifics of each case under consideration. However, in economic life there is a body of useful knowledge that cannot be subsumed under general rules. Here, the significance and frequency of changes in economic conditions are of paramount importance.[6]

Economic problems generally arise because of change in economic conditions [Hayek, 1948: 82]. Even more important, problems of change are central to economics. As Hicks [1979: xi] emphasizes, "[t]he more characteristic economic problems are problems of change, of growth and retrogression, and of fluctuation." There can be little doubt that change has become increasingly important in economic affairs. As increasing importance is attached to change as such, knowledge of particular circumstances of time and place becomes more valuable. This knowledge is essentially subjective, as is the process of learning involving changes in knowledge. In Kirzner's [1978: 35] words, "[a]ccording to the subjectivist point of view, economic change arises not so much from the *circumstances* relevant at the moment of decision, but from man's *awareness* of these circumstance." It follows that the ultimate decisions should be left to those most familiar with them [Hayek, 1948: 83–84].

Considerations of time and place are of course especially relevant to capital goods in general, and real property in particular. Buildings are durable and immobile. Land is permanent and immobile, whereas land uses may change over time. The enormous amount of detail inherent in the building process as a whole precludes unified statistical treatment. Statistical aggregates are considerably more stable than the underlying details, and generally do not show when it is necessary to respond to changing conditions. This applies to both opportunities for profit and threats of loss throughout the building process. Of course, not all aspects of the building process fall outside systematic control. For example, it will be shown in Chap. 5 that statistical tools are of great potential value in some aspects of real property management.

The network of individuals participating in all phases and facets of the building process cannot be replaced by a central office concerned with real property in a private or public organization without a significant loss of relevant knowledge. This does not mean that the coordination of the decentralized planning process underlying such a network cannot be improved. However, the essential task is to provide the coordinating function while decentralizing decision making sufficiently to prevent the loss of knowledge borne by every individual involved. Effective management of the building process requires that each participating individual has an adequate range of unhampered action, within which to take the responsibility for all decisions.

REAL PROPERTY AND GOVERNMENT INTERVENTION

Property is defined as the exclusive right to possess, enjoy, and dispose of a thing. By implication, this right affects only the interests of the owner in terms of both costs and benefits. Real property denotes the ownership in buildings and land. The specificities of real property demand special attention. Mechanically extending the conception of property as the exclusive right of ownership from mobile things to real property harbors a serious danger. In Hayek's [1948: 113] words:

> Where the law of property is concerned, it is not difficult to see that the simple rules which are adequate to ordinary mobile "things" or "chattel" are not suitable for indefinite extension. We need only turn to the problems which arise in connection with land, particularly with regard to urban land in modern large towns, in order to realize that a conception of property which is based on the assumption that the use of a particular item of property affects only the interests of its owner breaks down. There can be no doubt that a good many, at least, of the problems with which the modern town planner is concerned are genuine problems with which governments or local authorities are bound to concern themselves.

One of the important tasks of economics is to illuminate what may be legitimate and necessary government activities with respect to real property. The principles of private property and freedom of contract, essential to the functioning of the market process, require special attention in the case of real property.

In modern economic terminology, the favorable and unfavorable consequences of an action that extend beyond those affecting the economic agent in question are called external benefits and costs, respectively. The existence of externalities is interpreted as an imperfection in the market process, which consequently justifies government intervention. According to Mises [1966: 655]:

> Carried through consistently, the right of property would entitle the proprietor to claim all the advantages which the good's employment may generate on the one hand and would burden him with all the disadvantages resulting from its employment on the other hand. [. . .] But if some of the consequences of his action are outside of the sphere of the benefits he is entitled to reap and of the drawbacks that are put to his debit, he will not bother in his planning about *all* the effects of his action. He will disregard those benefits which do not increase his own satisfaction and those costs which do not burden him.

Control of such external effects is a proper function of the government. Zoning and building codes are classic examples of such interventions. One of the main reasons for government intervention in the building process is to increase the external benefits and reduce the external costs associated with real property decisions. Of course, this is more easily said than done. The long time-horizons involved in the estimation of external effects of a building project concern the uncertain future. Estimations of these effects may vary widely. They are intrinsically subject to interpretation, and thus to disagreement. We will return shortly to the subjective character of all economic valuations.

It should be borne in mind that the "division of labor" between private and public domains concerned with real property cannot be made once and for all. The interaction between the two domains unfolds in a social and therefore historical process. The very character of real property decisions, which generally have long-term effects, nevertheless demands a clear set of principles concerning private ownership in buildings and land at any moment in time. In fact, it is precisely for this reason that the market in buildings and land would benefit most from a predictable government policy bound to an intelligible set of fixed rules, rather than a discretionary policy easily affected by the whims of the political process, or by the bewildering array of legal subtleties. This is perhaps the most important conclusion that we can reach at this stage of the argument.[7]

SUBJECTIVIST CONCEPTION OF COSTS

Economists and accountants, including cost estimators and quantity surveyors, perceive the world somewhat differently. *Ex ante* or forward-looking concepts predominate in economics, while *ex post* or backward-looking concepts are more prevalent in accounting. However, a host of *ex post* concepts still survive in economics. They are a legitimate concern of economic history, but they cannot play a major role with respect to the future. Forward-looking concepts are indispensable in the theory of capital and in the investment choices encountered in practice. All capital decisions depend on our expectations and preferences regarding the future. The future is unknown and unknowable, however. We must assume not only that the future is unknown to an outside observer of the capital formation process, but also that this assumption is shared with those who make investment decisions.[8]

Cost is often treated as the preeminent *ex post* concept. However, no matter how accurate and exhaustive our historical records, backward-looking concepts of cost are inadequate for two reasons. First, at the moment of decision one is perforce considering *future* costs. Second, the valuation of costs is impossible without an explicit account of opportunity costs—the satisfactions foregone. As we will see, concepts of opportunity cost and utility are inherently related. This aspect of cost valuation will be intractable no matter how good a prediction itself may be. Costs must remain in a world of projecting and imagining.[9]

The activities of projecting and imagining are free to the extent that the world is indeterminate. Imagining the future, including the future costs, is a creative endeavor. The future must be created rather than discovered [Shackle, 1969: 16].[10] The same applies to costs. In a sense, they are fully determined precisely at the moment when they cease to matter to a decision maker, that is, when the opportunities for choice are fully exhausted.

The difficulties with cost forecasting based on historical data are exacerbated in the case of long-lived capital goods, such as buildings. With respect to both costs and benefits, building decisions must be approached from the point of view of a decision maker who can make a difference, and whose actions *ipso facto* cannot be determined once and for all. This requires a subjectivist framework, in which costs and benefits become known as new knowledge is acquired. Of course, knowledge acquisition is also time-dependent. In Chap. 5, we will discuss in considerable detail some methodological aspects of this issue.

COST, CHOICE, AND INFLUENCE

As a building project unfolds from planning and design, through procurement and construction, to utilization and operation, and as the cumulative cost of the project rises, the opportunities for influencing the cumulative project cost diminish. This insight is often employed to emphasize the importance of early decisions upon the total project cost. More specifically, it is argued that the planning and design phase of a building project deserves more abundant resources and closer attention because in this phase we simultaneously encounter the greatest possibilities for influencing the total project cost and the lowest expenditures associated with the project.[11] The shrinking opportunities for economizing the use of building resources are illustrated diagrammatically in Fig. 3.1.

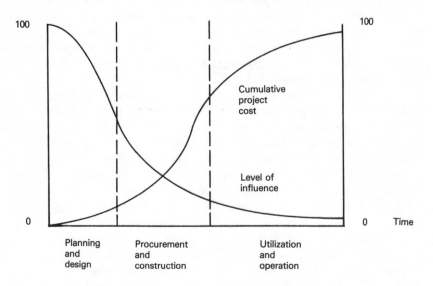

Figure 3.1.   Level of influence on project cost (in percent).

Of course, this argument is plausible on *a priori* grounds. In the initial project phase we should consider as many alternatives as possible for economizing the use of building resources. This argument is typically advanced only in terms of the project phases, rather than in terms of real time. Various building professionals often perceive the building process in this discontinuous fashion. However, it is true not only that the planning and design phase offers the greatest opportunities for economizing, but also that these opportunities shrink most dramatically in this project phase. A building project "ossifies" well before it reaches the construction site. Although the prospective building may still be in the form of graphite deposits on drafting paper, or electronic impulses on a CAD system's display terminal, it nevertheless assumes a reality of its own.

How can we explain this phenomenon? The sunk costs associated with planning and design are rarely the reason for this rapid ossification, especially from the vantage point of the client. Of course, these sunk costs may be a major concern of the professionals involved in building planning and design, who are very much constrained by the project budget and schedule.

The method and cost of financing may play a considerably more important role in pushing the project forward, despite the knowledge that further planning and design might be beneficial. For example, in the case of speculative building—when a building project is initiated by a developer—planning and design expenditures are covered by the developer's "out of pocket" money, whereas the construction expenditures are heavily leveraged. Although the planning and design expenditures may be minuscule in comparison with the total building cost, they tend to be significant with respect to the developer's own budget. However, the driving force behind the rush is still elsewhere—in the opportunities for profit foregone because of project delays.

The opportunity cost of project delays is ultimately associated with the postponement of project delivery, that is, in the delay of the useful services the client expects the building to provide. The argument concerning total project cost is misguided to the extent that it neglects these costs in terms of opportunities foregone from the vantage point of the client. Even more important, the opportunity cost of a project is greatest at the onset. It is reckoned in terms of all other alternatives foregone because of that project. Broadly speaking, the opportunity cost of a project, as perceived by the client, follows the same pattern as the level of influence on total project cost: it declines as the project unfolds. This is the main reason behind the customarily hectic pace of the building process in its early stages.

OPPORTUNITY COSTS, SUNK COSTS, AND CONSTRUCTION COSTS

When we are considering an investment decision, the capital budgeting and benefit-cost literature advises us that we should disregard all sunk costs. We have already encountered this argument in Chap. 1. Bygones are forever bygones. An investment project should be undertaken only if it more than pays for itself. In short, we are instructed that it is economically unsound to attempt "correcting" past investment mistakes by new investment expenditures. Our decisions cannot affect the past, but only the future. The alternatives open at the moment of choice are always in some essential sense perceived subjectively by the decision maker.[12]

This argument applies to both new and replacement investment decisions. An investment project involving replacement of a worn out or technologically obsolete component of an existing capital asset must also more than pay for itself, irrespective of the costs sunk in that capital asset in the past. For example, a leaking roof will be neglected—or inexpensively patched up—rather than replaced if the replacement expenditures cannot be justified by the expected return from the economic process affected by the roof leak, even though the building in question may have already absorbed considerable resources. Therefore, under some conditions we may be best advised to abandon an existing capital asset, that is, to disinvest. Disinvestment is always among the alternatives present at the moment of choice. Investment and disinvestment decisions can be made with the same set of capital budgeting decision criteria: we should abandon the capital good, or its component, which has the least present value.

Of course, all this pertains to building investment decisions in general. In fact, most investment projects require some building, due to the impossibility of unbundling various components of the so-called plant and equipment. However, the completion of major investment projects typically requires a rather long period of time. The longer the construction period, the more likely it is that economic conditions will change, thus changing the economic viability of the project. The construction period therefore affords many opportunities to reconsider the project as a whole, well before it reaches completion. Clearly, the argument concerning sunk costs can be invoked at any point in time, and not only upon completion of a project.

The subjectivist conception of cost, with its emphasis on an individual's perception of the opportunity cost of each available course of action, is essential in this context. In this view, cost is the prospective opportunity displaced by the decision to take one course of action rather than another. Costs must be reckoned in terms of satisfactions foregone.[13] Somewhat more technically, in a theory of choice a cost must be reckoned in a utility dimension, rather than a commodity dimension [Buchanan, 1969: 43]. Cost thus represents the anticipated utility loss associated with a rejected alternative.

As a construction project unfolds, funds are committed to the project in phases. It should be borne in mind, however, that each phase in this process is perceived by the client, the decision maker who is undertaking the investment project, in the light of ever new opportunities for profit foregone because of the project. In a sense, the sunk cost portion of the project grows during the construction period, while the remaining choices gradually shrink and in some cases ultimately vanish. In fact, many construction projects are abandoned in mid-course because the client feels the "remaining" opportunity cost is too great. Put differently, bygones are forever bygones even during the construction process. Therefore, it may be argued that there will be an incentive for both owners and builders to shorten the construction period. We will return to this topic in Chap. 4 in the discussion of building technology.

CONSTRUCTION PRODUCTIVITY
AND QUALITY OF CONSTRUCTION GOODS

The construction sector is often summarily dismissed as the most backward sector in the economy. Both from a technical and an economic standpoint, this sector is regarded as stagnant.[14] Conventional productivity measures, that contrast total factor inputs and the resulting output, support this claim. People acquainted with the industry know that construction technology has, in fact, changed a great deal over time. In a sense, a host of technological innovations are carefully hidden from view due to the proverbial conservatism of construction markets. Nevertheless, the problem should be approached by concentrating on the underlying economic issues, that is, by focusing on the specific character of the construction product rather than on construction technology as such [Rosefielde and Mills, 1979: 93]. In particular, it is important to bear in mind the differences between construction goods and mass-produced goods, broached in Chap. 1.

Construction goods are immobile and durable. These characteristics require them to be functionally flexible. According to Rosefielde and Mills [1979: 93–94]:

> Most construction durables are locationally immobile. Unlike ordinary mass-produced goods, they cannot be used interchangeably or moved about at will. As a consequence of this immobility, construction durables are usually designed not only to achieve some particular function at some specific site but also to meet the constraints and variable opportunities afforded by changing economic conditions. Good construction design therefore stresses functional flexibility. It facilitates the production of diverse outputs and the employment of capital and labor over a broad range of combinations of inputs in various proportions. Moreover, it governs the durability of the construction good itself, since durability is a crucial element of long-run marginal cost.

The interdependence between design and construction is especially important in this connection. Most buildings that serve as production goods are custom-designed and built. Their design reflects specific circumstances of time and place, as perceived by the client and the team of building professionals the client assembles for the occasion. In short, buildings are heterogeneous capital goods, discussed in Chap. 2. Many of their qualitative characteristics are not captured by standard productivity measures, which focus on quantitative growth.[15]

As a consequence, any calculation of construction productivity that fails to account for change in construction quality is likely to result in an underestimation of technical progress. Conventional productivity measures are therefore likely to be of limited value in the case of the construction sector. Factors other than the volume of output may be combined under the notion of product quality. The apparent stagnation of construction productivity may be due to the neglect of these factors [Rosefielde and Mills, 1979: 94].

The productivity issue is of particular concern to us here because the standard view of construction goods neglects their flexibility and adaptability. This view disregards the essential characteristics of the building process. Given the length of the construction period, construction goods must be designed, constructed, and operated with functional flexibility in mind. Flexibility and adaptability should not only be characteristic of completed construction goods: throughout the building process a building may undergo significant changes. As we will see later on in this chapter, the owner often changes the design of a building even in the construction phase. The question of quality revolves around this essential feature of construction goods. It goes without saying that building quality cannot be reduced to flexibility and adaptability only.[16]

FIXED CAPITAL, CIRCULATING CAPITAL, AND LIFE-CYCLE COSTING

In the conventional treatment of capital goods, stress is laid almost exclusively on the role of fixed capital, as though capital consisted only of very durable goods. In a theoretical framework focusing on the use of capital goods, the role of circulating or working capital is naturally of paramount importance.[17]

This is not a mere academic trifle. Investment decisions in practice tend to exhibit the same bias as does conventional capital theory. Managers involved in investment decisions tend to think of investment projects exclusively in terms of fixed assets, and they often neglect the long-term implications of a project in terms of circulating or working capital. According to Arnold [1986: 117], for example:

> Managers without a financial orientation often ignore or dramatically underestimate the total amount of money a project ultimately requires. They think in terms of fixed assets—land, building, and equipment—and do not think enough about the additional investment in net working capital.

The life-cycle costing literature of course pays special attention to the total cost of ownership of buildings. However, in actual practice these costs are rarely estimated beyond the costs explicitly related to the building itself, such as maintenance, replacement, and operation costs. The so-called functional-use costs, composed primarily of the labor and material costs associated with the function of the building, are almost never estimated in any detail. Functional-use costs associated with many building types represent a very high proportion of total costs. For example, the functional-use costs are believed to be between 60 and 90 percent of the total life-cycle costs of a wide range of building types such as offices, hospitals, and universities.[18]

It should be borne in mind that functional-use costs generally overlap with the notion of working or circulating capital. This is important in view of the terminological differences that exist between fields such as building economics and investment finance.

Although it may be exceedingly difficult to evaluate the functional-use costs of buildings, one of the research tasks facing us is twofold. First, the key components of functional-use costs should be identified, and second, their relative magnitudes (or at least orders of magnitude) should be identified for different building types. However, before we can approach this research task we must address the underlying conceptual issues, which concern the relationship between fixed and circulating capital assets. This is one of the reasons why it is useful to think of all capital in terms of circulating or working capital.

MAINTENANCE AND REPLACEMENT
OF BUILDING COMPONENTS

As we have already seen, capital maintenance and replacement cannot be regarded as routine activities. They require continual attention, especially for long-lived capital assets. There are two types of problems regarding capital maintenance and replacement. First, the determination of the "best" levels of maintenance and replacement, and second, the solution of problems arising in the process of carrying out such decisions [Lachmann, 1986: 69].

Problems of the first kind concern the preservation of an income stream over the relevant time horizon. This time horizon depends on the plan informing the capital combination in question. The preservation of a desired income stream requires decisions about the time-profile of the corresponding maintenance and replacement activities. Expectations concerning both time sequences of income and expenditures for maintenance and replacement dominate this problem.[19]

Problems of the second kind arise whenever gross profits turn out to be inadequate to finance the planned expenditures for maintenance and replacement. A revision of such plans is then required. The revised plans may fail, too. Of course, maintenance and replacement plans may not fail, but we should recognize their problematic character.

Let us consider maintenance and replacement expenditures in some detail. It is useful to distinguish between expenditures for maintenance *sensu stricto,* including minor repairs, and for replacement. The former are typically covered out of the current account, while the latter draw on the capital account. Decisions involving current and capital account expenditures are often made by individuals from different parts of an organization. Furthermore, maintenance and minor repair work is typically performed by "in-house" crews, while replacement of building components often entails the services of a specialized contractor. Although the two types of expenditures undoubtedly share a fuzzy boundary, the fact that replacement expenditures are commonly treated as investment expenditures is important in building economics.[20]

We should bear in mind throughout this discussion that building component replacement is an explicit or implicit aspect of building design. We may regard buildings as combinations of heterogeneous building components, conceptually analogous to capital combinations formed by different capital goods. Many of these building components are indeed regarded as capital goods requiring investment decisions. For example, roof, elevator, or boiler replacement is such a decision, which may have to be made several times during the building life cycle. Even more important for our present purposes, these building components may be replaced by a wide variety of near substitutes. Issues of complementarity and substitutability, analyzed in Chap. 2, are therefore relevant to various components of individual buildings, as well. Building design concerns both complementarity and *ex ante* substitutability of building components, whereas building management also concerns *ex post* substitutability. Architectural and engineering designers must provide for the future decoupling of building components needing replacement without major interference with building utilization and operation. In this context, they must consider both physical and economic implications of building component decoupling. The expected life cycle of a building component, together with its interaction with the life cycles of related building components, is therefore one of the most important considerations in its design or selection.[21]

Returning to our argument, and disregarding for the moment the problem of determining the desired income stream associated with a building serving as a production asset, it is evident that there is a trade-off between maintenance and replacement costs. An increase in the level of maintenance of a particular building component will result in a decrease of its replacement cost per unit of time, other things remaining equal. In other words, improved maintenance will tend to increase the expected useful life of a building component. As the level of maintenance increases, the maintenance cost monotonically increases, while the replacement cost monotonically decreases. By the same token, the total cost of maintenance and replacement will have a minimum, indicating the "best" level of maintenance [Bon, 1988]. This is illustrated in Fig. 3.2 (note that all costs should be expressed in time-equivalent currency units). According to this argument, a particular building component may be either under- or over-maintained. This is of conceptual interest because it is often assumed that more maintenance is always better.

However, it should be noted that this argument concentrates exclusively on costs. The relationship between benefits and costs of maintenance and replace-

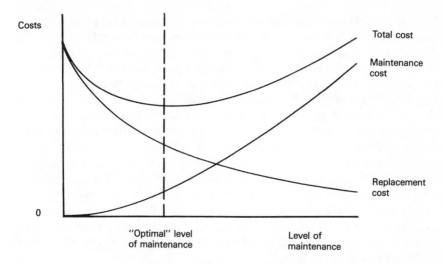

Figure 3.2. "Optimal" level of maintenance.

ment ultimately depends on the income stream expected from the entire building as a productive asset. The above argument implicitly assumes that the income stream does not vary over time.

With this proviso, we can understand a sequence of circumstances that may lead to a failure in maintenance and replacement plans. If gross profits are inadequate to support the planned maintenance expenditures, the expected useful life of a particular building component would decrease. Deferred maintenance eventually leads to the need for much more costly replacement. If the same condition pertaining to gross profits persists, the building component would not be replaced once it failed. The failure of one building component is likely to result in rapid deterioration and ultimate failure of other components. A prolonged period of deferred maintenance would ultimately lead to a need to replace the entire building. For example, progressive building deterioration is often associated with water penetration, due to the initial failure of roofing or exterior cladding. The process unfolds rather rapidly if it is not checked by repair and replacement activities.

We should note in passing that the economic life of a replaceable building component should be determined on the same basis as that of any other investment project. As was argued in Chap. 1, it must be more profitable to carry on up to the economic horizon than to any other time-horizon. In this case, the economic horizon refers to the "best" replacement cycle of a building component. Again, the test of profitability is the present value calculation.[22] For example, a roof should be replaced every 20 years if the present value of this replacement cycle is greater than the present value of any other alternative cycle.

THE USE VALUE AND EXCHANGE VALUE OF BUILDINGS

Economic goods may have both use value and exchange value to their owners. This applies to buildings, as well. Economic goods can be employed either for direct or indirect satisfaction of our needs. Although both forms of employment of economic goods are subject to the same general principle of value determination, in some cases it may be important to make a clear distinction between the two, as well as to keep track of their relationship.

The use value and exchange value of a good, as two forms of value, are often of different magnitudes. The question that arises is which of the two is the economic form of value in a particular case [Menger, 1981: 230]. In all economic activity, the actors seek the fullest possible satisfaction of their needs. This follows directly from the consideration of the nature of value. Therefore, whenever a good has both use value and exchange value, the *economic* value is the one that is greater in magnitude [Menger, 1981: 230–31]. At any given time, an economizing individual will weigh the relative importance of the two forms of value. The decision to keep or sell a particular good, or a portion of the good, will depend on this evaluation.[23]

The interdependence of use value and exchange value is often neglected in business practice. The two forms of value are often regarded as independent. The primary emphasis is placed on the specific motivation behind an investment decision, either use or exchange. This motivation thus implicitly determines the dominant form of value, while the relationship between the two is abstracted away until the moment the investor's motivation changes.

Consider a related example, concerning the value of real property. Why is real property valuable? According to Epley and Millar [1984: 6], real property can be classified into two categories depending on whether a buyer intends to use it or to sell it at a later date. The value in the former case is determined by use, while in the latter it is determined by exchange.

Their argument runs as follows. Both use and exchange values are demand-oriented. On the one hand, buyers who purchase real property in order to sell it are primarily motivated by the expected price appreciation, that is, capital gain. The property is purchased because it is expected to be attractive to others. The property in question may or may not be producing income currently, but income is secondary in importance to capital gain. On the other hand, buyers who purchase real property in order to use it are primarily motivated by the income stream produced by it. A market for the product of the property is necessary to create an income stream large enough to make the investment attractive in comparison with alternative investment opportunities.

This is where Epley and Millar leave the subject. However, the fact is that use value and exchange value are deeply intertwined. All speculation implicit in the exchange value is based on the expectation that buying and selling transactions will ultimately cease and someone will buy the property for use. In fact, all speculative buying and selling is based on the expectation of capital gain precisely in relation to the potential uses of the property. In the last analysis, the expected use value dominates the exchange value. Undoubtedly, the two types of value affect each other. A buyer intending to use a piece of property will also think about its resale value. If a venture is risky, a contingency plan might involve selling the property before the end of the investor's planning horizon. Thus, the exchange value will play an important role in a buyer's plan even when his or her primary motivation is to use the property.

The complex interaction between the two forms of value goes even deeper. Speculative buying and selling may drive up the price of property, and this price may then determine the use by setting the lower limit on the income stream that could be obtained from the products it helps produce. The speculative bubble may burst, leaving the last owner with a property the exchange value of which must be revised downward drastically. The property would thus become accessible to less profitable uses.

Of course, the relationship between use value and exchange value of a good is by no means static. It changes with changing economic conditions. An economizing individual should therefore keep track of the direction of change of the use value and exchange value of an economic good. Clearly, anything that diminishes the use value of a good may, other things being equal, cause the exchange value of the good to become the economic form of value, and vice versa [Menger, 1981: 231]. Therefore, one of the tasks of economizing is to identify the causes that may change the relationship between the two forms of value. There are three general causes of changes in the economic form of value. First, changes in the owner's preference structure; second, changes in the properties of a good, and third, changes in the quantity of a good in question [Menger, 1981: 231–33].[24]

Obviously, there is no difficulty in extending this argument to durable consumption goods including buildings. More important for our purposes, this argument can be extended to buildings as durable production goods, subject to the conditions already discussed in Chap. 2 concerning complementary goods of higher order. Namely, the value of any good of higher order, including capital goods, is determined by the value of goods of lower order that would remain unproduced if we were to lose the services of the higher order good in question. Neither land nor capital differ from other goods in this regard [Menger, 1981: 165, 172].

Consider, for example, the options available with owned properties, illustrated diagrammatically in Fig. 3.3. Qualitative reasoning formalisms, discussed in greater detail in Chap. 5, can be used by the decision maker to deduce which properties to watch, keep, or dispose of, given his or her assumptions about use value and exchange value trends concerning these properties. The opportunity cost of keeping a property should be included in its exchange value. Cost of operations can be added as the third dimension of this diagram. Note that Fig. 3.3 shows a particular combination of these trends, whereas both use value and exchange value can be increasing, stable, or decreasing. A similar

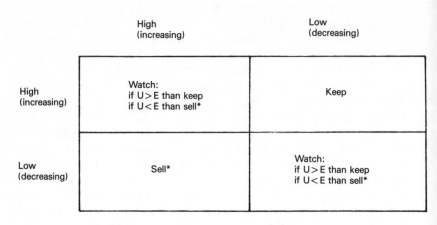

Figure 3.3. Options available with owned properties.

diagram can be constructed for properties potentially interesting for acquisition or leasing in the case where increasing needs for space are expected.

The implicit argument here is that the economic form of value of a particular good should determine whether it should be used or sold. As we have already seen, this relationship is, in fact, rarely observed in business practice. Namely, the "economic value" of a good is determined on the basis of a prior decision concerning either its use or its sale. The use-sale trade-off is considered only in a crisis.

In this context it is instructive to return briefly to the topic of capital accounting, discussed in the previous chapter. As we have already seen, in the case of long-lived capital goods, such as buildings, accounting valuations involve considerable guesswork. The valuation of a long-lived capital asset will depend on its prospective purpose or use. Its value will be assessed quite differently if the intention is to dispose of it than if it is expected to remain in operation. By implication, the intended purpose or use of the asset will determine its book value. Samuelson [1967: 95–96], for example, writes:

> Cash is the only asset whose value is exact rather than an estimate. All other valuations involve some guesswork, albeit careful guesswork. Moreover, all accounting valuations must be made relative to the actual intended purpose or use of the asset in question. If a business is a going concern and not in the process of liquidation, the accountant will be careful not to value doubtful assets at the low figure they would bring at a forced sale; he rather will value them at their worth to the company in its normal operation.

Of course, in a crisis the "low figure" an asset would bring at a forced sale might be higher than the asset's worth to the company.[25] We will return to this subject in Chap. 5, where we will focus on the need to continually monitor the real property operated by an organization.

THE BUSINESS CYCLE AND THE BUILDING PROCESS

How will various phases of the building process be affected by various stages of the business cycle? Here we will consider briefly the slump following a strong or a weak boom.[26] A strong boom is characterized by a capital goods shortage, a weak boom by insufficient demand. Capital regrouping, with its repercussions on maintenance and replacement, is central in both cases. In terms of the building process, the most interesting part of the story can be told in terms of the operation and utilization phase of the building process. Dealing with issues such as these requires a microeconomic perspective of the underlying economic process.

Well-informed actors in a market demonstrating a strong boom will sooner or later have misgivings about future yields and the cost of present projects. Alternatively, a rising rate of interest will dampen excessive optimism through the reinforced discounting factor. Capital reshuffling will then become necessary. However, capital reshuffling may also become necessary after a weak boom because of the appearance of excess capacity.[27] Under these conditions the firm may decide not to maintain a part of its capital, or not to replace it as it wears out. This will also depend on the control structure of the firm, discussed in the preceding chapter. The creditors of the firm may compel their debtors to dispose of a part of its capital.[28]

Here, we are dealing with the interaction between the plan structure and control structure, and indirectly with portfolio issues. The same thing happens during a slump following either a strong or a weak boom. In both cases capital regrouping may be required, which will affect the building process. A more stringent maintenance and replacement policy may affect all durable capital goods. In addition, capital goods in the phases of design or construction may be postponed or abandoned altogether. Finally, some durable capital goods may be sold if they cannot be fitted into revised plans.

Building maintenance and replacement policies should be pursued as active responses to economic change, rather than as "value preservation." Differential policies of maintenance and replacement for individual buildings also can be explained in terms of capital regrouping. More specifically, buildings that fit new plans better than others warrant a higher level of maintenance and replacement expenditures. Furthermore, some buildings may be converted to new uses at a lower cost than others, other things being equal.

Again, because building planning, design, and construction take so long, the process may be stopped at any point if business conditions change. A building may also be redesigned or reconstructed before it is put into operation if the project is continued under new business conditions.

The latter case is perhaps more interesting than the former, both in terms of its effects and in terms of its occurrence in the business cycle, that is, its timing. Many building failures (mostly failures in building performance) can be associated with a "change in scope" of the project during planning, design, or construction. By definition, change in scope is owner-generated. It is usually associated with a change in building function. The later the change occurs, the more significant its impact on future building performance. Even the owner-generated "change orders," smaller in scope, can be understood in this light. Their cumulative effect may be as significant as that of a change in scope.

The entire building process can be thought of as a mechanism for the transmission of knowledge about business conditions through the network of participating economic agents. The changes in business conditions will reflect the stage of the business cycle. For example, the behavior of the designer and the contractor will be influenced by their backlogs of projects and the prospect of new projects for current clients. In the early stages of a boom the backlog will increase, together with the likelihood of "repeat business." The most important indicators of business conditions will be propagated through the quantities and prices of building materials, design and contracting services, etc. These early warnings of an expected slump will be transmitted to the owner through the building process itself. In general, these warnings will occur before other signs of an impending change in business conditions, with the exception of rising interest rates.[29] We will examine these issues in greater detail in Chap. 4.

CONCLUSION

Although buildings and the built environment appear to change little over time, they are in fact continually adjusted to the changing economic conditions faced by the building owner. Maintenance and replacement activities are among the main vehicles of change. Contrary to the customary view that these activities "preserve" the building value, as well as the external and internal form of buildings, they represent mechanisms of continual adjustment.

Of course, the immobility and rigidity of buildings and land use patterns resist rapid adjustment. That resistance is therefore one of the key problems of building design and the underlying building technology. As the opportunity for change declines most in the early phases of the building process, the requirement for flexibility and adaptability of buildings is often most pronounced in these phases. The building owner will perceive these opportunities most clearly while the building is beginning to take shape. For the same reason, the entire building project will tend to be most vulnerable in the planning and design phase, when the opportunity for discontinuing the project may attract the owner most.

Thus far, we have emphasized the effect of change in economic conditions on the building process. The last section provides the foundation for Chap. 4, where we will consider some systematic aspects of economic change. In particular, we will examine the interaction between business and building cycles when there are systematic changes in the availability of credit, that is, loanable funds for investment purposes. In other words, we are now ready to proceed to an analysis of some systematic underpinnings of economic processes that affect the building process as a whole.

NOTES

1. According to Böhm-Bawerk [1959b: 7]:

> If we observe more closely how man assists the natural processes, we shall find that his sole but completely adequate activity lies in spatial control of matter. The ability to move matter is the key to all man's success in production, to all his mastery over nature and her forces.

> Mises [1980: 97] agrees: "No human act of production amounts to more than altering the position of things in space and leaving the rest to Nature." Both credit this idea to Mill [1973].

2. As Böhm-Bawerk [1959b: 8] writes:

> This suggests how man acquires control over the temporal point at which a given result appears. He need merely avail himself of his capacity for spatial transfer of matter with sufficient skill to assemble, by way of preparation, the causative factors of the desired result *with one exception*. [. . .] Now at the proper moment he brings his last partial or contributing cause into place, the delayed activity is suddenly released, and the desired effect is garnered at the appropriate time.

3. According to Böhm-Bawerk [1959b: 81], nature and land are the *technical* elements of production, while land and labor are its *economic* elements.

4. It should be noted, however, that Mises [1966: 492-93] and Hayek [1975: 119] argue that capital is not a factor of production, the quantity of which is given independently of human action. As we will see in Chap. 4, the quantity of capital is not given even in the short run, because it depends on capital maintenance, which in turn depends on entrepreneurial foresight.

5. According to Böhm-Bawerk [1959b: 166]:

> The theory of the value of complementary goods offers a solution for one of the most important as well as one of the most difficult problems in economics. I refer to the problem of the distribution of goods as it takes place under the present organization of society, in which more or less free competition prevails and prices are determined by contractual agreement. All products arise from the cooperation of

the three complementary "factors of production," labor, land, and capital. Our theory tells us how much of the combined product is economically attributable to each, and therefore how much of the integral value of the product is to be ascribed to each one.

For a related discussion, see notes 24-6 in Chap. 2.

6. As Hayek [1948: 80] argues:

> Today it is almost a heresy to suggest that scientific knowledge is not the sum of all knowledge. But a little reflection will show that there is beyond question a body of very important but unorganized knowledge which cannot possibly be called scientific in the sense of knowledge of general rules: the knowledge of the particular circumstances of time and place. It is with respect to this that practically every individual has some advantage over all others because he possesses unique information of which beneficial use might be made, but of which use can be made only if the decisions depending on it are left to him or are made with his active cooperation.

In this context Hayek [1948: 80] specifically mentions the real estate agent, "whose whole knowledge is almost exclusively one of temporary opportunities." Although Hayek's remark about the sacrosanct nature of scientific knowledge strikes us as fairly outdated today, we will see in Chap. 5 that this form of scientific orthodoxy is far from defunct.

7. New York City offers a good example of a real estate "market" dominated by the political process. As Tuccille [1985: 234], for example, writes, "[w]hy not let market conditions determine what is or is not needed instead of trying to force or hasten change by subsidizing a privileged few?" Donald J. Trump is one of the privileged few. In an environment choked by rules and disputes about their interpretation, a remarkable and resourceful individual like Trump can move the rules in his favor, thus internalizing most benefits and externalizing most costs of development. The surfeit of rules—an index of the distrust of the market process—unwittingly leads to the condition it was supposed to circumvent.

8. As Hicks [1977: vii] observes, "one must assume that the people in one's models do not know what is going to happen, and know that they do not know what is going to happen."

9. As O'Driscoll and Rizzo [1985: 47–48] write:

> The only sense in which cost can influence choice is the perception at the very moment of choice of the satisfactions foregone [. . .]. Thus, cost is tied to choice and apart from it it has no economic meaning [. . .]. After a choice is made, retrospective calculations of what the relevant costs "really" were (in the sense of what the actor would have perceived if he had certain superior knowledge) cannot, of course, be relevant to that prior choice situation. To the extent that conditions are expected to remain the same, however, such calculations can inform future choice situations. Nevertheless, even these historical estimates cannot demonstrate unambiguously what *would* have happened if the individual had chosen another

course of action [. . .]. Thus, even *ex post,* costs cannot be realized. They must remain forever in a world of projecting, fantasizing or imagining.

For a methodological discussion of the subjectivist conception of costs in the context of building economics, see Bon [1986b and 1987].

10. As we have seen in Chap. 1, "[t]he future is unknowable but not unimaginable" [Lachmann, 1978b: 3]. Lachmann [1982: 38] credits this insight to Mises.

11. See Paulson [1981: 422–26], for example. Figure 3.1 is based on Barrie and Paulson [1978: 154], also reproduced by Paulson [1981: 423].

12. As Buchanan [1969: 16–17] writes:

> At any moment in time, one can look either forward or backward. One looks backward in time in a perspective of foreclosed alternatives. One looks forward in time in a perspective of alternatives that still remain open; choices can be and must be made. With this elementary clarification, cost tends to be a part of choice among alternatives, a choice that must be subjective to the chooser.

13. According to Buchanan [1969: 28], "[a]ny profit opportunity that is within the realm of possibility but which is rejected becomes a cost of undertaking the preferred course of action."

14. Rosefielde and Mills [1979: 83], for example, write:

> Some economists claim that little technological change occurs in the construction industry. It is often said that buildings are constructed the same way they were centuries ago, and that construction is a handicraft industry. Labor productivity is said to be too low, and the industry is generally held to be costly and technologically stagnant.

15. As Rosefielde and Mills [1979: 94] write:

> All these considerations taken together imply that construction durables are almost inevitably heterogeneous to the extent that they effectively use relevant locational opportunities. The measurement of growth and productivity in the construction sector therefore poses special difficulties. Quantitative calculation of the numerical increase in growth and productivity tell us very little about their qualitative growth; growth in quality is expressed by the contribution these durables make to the final volume of goods and services produced elsewhere in the economy.

16. In a comprehensive survey of quality attributes of architecturally significant buildings in Great Britain, Powell [1987: 29] shows that building flexibility is, in fact, regarded by those surveyed as the least important quality attribute. Perhaps predictably, given the nature of buildings in the sample, the survey shows that "soft" attributes, such as building appearance, dominate "hard" attributes, such as building life cycle performance. It is interesting to note that Powell [1987: 32] concludes his article with a question: "[is this] further evidence of British indifference to the values of production and economic success?"

17. For Hayek's [1941: 47–49] comparison of these two perspectives, designated as "Anglo-American" and "Austrian," see the Appendix in Chap. 2.

18. For a detailed discussion of functional-use cost estimates, see Ward [1987], for example.

19. On this problem, see Hayek [1975: 83–134]. This paper, entitled "The Maintenance of Capital," was reprinted from *Economica,* Vol. II (New Series), August 1935.

20. Lachmann [1986: 69] subsumes maintenance, repair, and replacement under the general notion of "maintenance and replacement." This is too broad an aggregate for our present purposes. In fact, it may be argued that these activities generally should not be subsumed under an aggregate at all. Maintenance and repair expenditures, on the one hand, and replacement expenditures, on the other, are reflected in current and capital accounts. This distinction—customary in business practice—indicates the essential difference between these two types of activities.

21. For a discussion of related design issues, especially with regard to housing, see Habraken [1983], for example. Tempelmans Plat [1982 and 1984] provides an economic framework for Habraken's approach to building design. It should be noted that both emphasize the importance of building utilization and operation as the basis for determination of building component durability. Put differently, durability issues should be approached from the vantage point of the building users' needs.

22. For an example of a problem involving replacement timing with rising maintenance costs, see Baumol [1977: 611–13]. For a general framework concerning maintenance and replacement activities in organizations with large real property portfolios, see Bon [1988].

23. As Menger [1981: 231] writes:

> One of the most important tasks of economizing men is that of recognizing the economic value of goods—that is, of being clear at all times whether their use value or their exchange value is the economic value. The determination of which goods or what portions of them are to be retained and which it is in one's best economic interest to offer for sale depends on this knowledge.

24. Menger [1981: 233] argues that the third case is perhaps the most important of the three: "An increase in the quantity of a good a person has almost always, other things remaining the same, causes the use value of each unit of a good to him to diminish and its exchange value to become the more important." The converse is true, as well. Menger [1981: 234] also mentions that the effect of changes in total wealth of an economizing individual are of special importance in this connection.

25. For a related discussion, see notes 28, 30, and 31 in Chap. 2.

26. The most interesting part of the Austrian business cycle theory, as presented by Lachmann [1978a], deals with this particular sequence of events. According to Lachmann [1978a: 113], the Austrian theory is a theory of a strong boom.

27. According to Lachmann [1978a: 125]:

> Capital regrouping is thus the necessary corrective for the maladjustment engendered by a strong boom, but its scope is not confined to this kind of maladjustment. Where a weak boom has "petered out" before "hitting the ceiling" capital

regrouping is just as necessary. That this is not at once obvious is due to the unfortunate habit of viewing all these problems in "macro-economic" terms.

The point that capital regrouping may be necessary in the wake of both strong and weak booms is important because the Austrian theory emphasizes the strong boom case. Lachmann [1978a: 125] continues:

At the end of a weak boom the new capital resources begin to pour out output. Here there is no difficulty in obtaining and keeping together the complementary factor combinations since the ceiling has not been hit. But some firms may find it difficult to dispose of the new output. Prices will tend to fall, employment may decline and unsold stocks accumulate. Excess capacity (of the "real kind"!) may make its appearance.

28. As Lachmann [1978a: 121] argues:

The owners of a factory are unlikely to close it and let their plant idle merely because their liquid assets could earn a higher rate of interest elsewhere. But here the control structure is of some importance. In certain cases the creditors may compel their reluctant debtors to part with their mobile assets. In general we may say that where the division between the firm's own capital and its debts corresponds most closely to that between its first-line, and its second-line and reserve assets, i.e., its fixed and mobile resources, the chances of a successful withdrawal of the mobile resources are highest, precisely because all the gain will accrue to the creditors and all the loss to the debtors. While, where the creditors own also a part of the fixed first-line assets, they may be reluctant to incur the capital loss here involved. But in any case there will be enough resistance to all attempts to mobilize resources and disintegrate existing combinations to make the withdrawal of mobile factors a slow and precarious business.

29. It is interesting to note that changes in commercial and industrial building contracts have been included in the so-called leading economic indicators of business cycle peaks and troughs since the inception of their systematic monitoring in late 1930s. For an historical overview of the leading economic indicators, see Moore [1983: Chap. 24].

4

Business
and
Building Cycles

INTRODUCTION

The building process is affected by economic change like any other economic process. However, long-lived capital goods, such as buildings, are affected by these changes to a greater extent than short-lived capital goods. As we argued in Chap. 1, time matters more in the former than in the latter case. In this chapter, we will explore the effect of systematic changes in economic conditions on the building realm. In particular, we will concentrate on fluctuations in building activity stemming from systematic preference changes.

Our attention will focus on the interaction between business and building cycles. We should point out, however, that building cycles considered here are not those associated with the work of Kuznets [1961], which are considered to be 15 to 25 years in length. Although the Kuznets cycles are of great historical interest, there is no reason to expect that they will persist in the future. Instead, we will consider the building activity in the context of business cycles with durations of three to five years.

Our presentation will proceed from general questions about economic fluctuations to fluctuations in building activity, that is, the relations between business and building cycles. For this reason, the introductory sections will be rather abstract. First, we will consider three types of preference changes with economy-wide consequences. They are liquidity, leisure, and time preference changes. Here, we will focus on time preference changes in the absence of policy disturbances affecting credit availability. Second, we will discuss the consequences of a policy-induced expansion of credit on investment behavior. Special emphasis will be placed on the entrepreneurial decisions concerning the distribution of short-lived and long-lived capital goods during the artificial upswing. Third, capital malinvestment in the context of policy-induced credit expansion will be explored in some detail. We will focus on both the abortive upswing and the ensuing contraction, which requires adjustments in the capital structure. Fourth, we will briefly discuss the learning process that accompanies capital malinvestment. We will argue that the cycle cannot be reproduced time after time, because people learn from their experience. Fifth, the building cycle will be examined in considerable detail, based on the preceding analysis. Again, we will emphasize the distinction between short-lived and long-lived capital goods. Sixth, we will return to the problem of capital maintenance and replacement in the context of technical change. Some trends in the development of building technology will be examined in light of the requirements of continually changing economic conditions.

PREFERENCE CHANGES, OUTPUT VARIATIONS,
   AND CAPITAL RESTRUCTURING

What causes the level and structure of output to vary over time in a smoothly functioning economy? This question is fundamental to macroeconomics. It focuses on preference changes that have a predictable influence on the level and structure of output. We are interested primarily in those preference changes of systematic and economy-wide consequence. Preference changes influencing only particular markets for goods and services lie in the realm of microeconomics.[1]

Before we proceed, a caveat is in order. In keeping with the level of generality of the entire book, we will disregard here the effect of changes in the foreign sector. Clearly, the importance of the foreign sector depends on the size of an economy: the larger the economy, the less important will it tend to be. The exclusion of the foreign sector from consideration will therefore render the discussion somewhat too general with respect to economies that depend substantially on international trade. However, it should be noted that the validity of the arguments presented here does not depend on our simplifying assumption.

Liquidity, leisure, and time preference changes are the three types of preference changes of special interest here. They are related directly to the key macroeconomic aggregates of money, employment, and saving and investment. Furthermore, alternative preference changes serve as keystones for alternative classes of macroeconomic theories. Changes in time preferences are of special concern to the Austrian school. We will discuss them in turn.[2]

First, the level of output may change in response to a change in liquidity preferences, that is, a change in the demand for money or real cash holdings. Underlying these preferences is the money-goods trade-off. An increase in liquidity preferences would result in an increase in the quantity of money demanded and supplied and a decrease in the quantity of all other goods, including capital goods.

Let us concentrate on capital goods. If we break down this broad aggregate, we may consider the relationship between short-lived and long-lived capital goods as liquidity preferences change. It is useful to think about the distinction between short-lived and long-lived capital goods as analogous to the distinction between circulating and fixed capital. We must be careful not to confound these two distinctions, however.[3]

Now, transaction balances, a major component of real cash holdings, and short-lived capital goods are complements, whereas transaction balances and long-lived capital goods are substitutes. As a consequence, an increase in the demand for real cash holdings would be accompanied by an increase in the demand for short-lived capital goods and a decrease in the demand for long-lived capital goods, and vice versa [Garrison, 1985: 173–74, following Kessel and Alchian, 1962: 534]. Therefore, a change in liquidity preferences would lead to systematic capital restructuring, that is, a systematic change in the time-structure of the process of production.[4]

Second, the level of output may change as a consequence of a change in the demand for leisure, that is, a change in the supply of labor. The leisure-labor trade-off in fact implies a leisure-goods trade-off. An increase in leisure preferences, that is, a decrease in the supply of labor, would result in a smaller quantity of other goods.

In this connection, it is also useful to distinguish between capital goods that are complements of labor from those that are substitutes for labor. A decrease in the supply of labor would result in a decrease in the demand for complementary capital and an increase in the demand for capital that serves as a substitute for labor, and vice versa. Thus, changes in leisure preferences will have a systematic effect on the structure of production.

Third, the level of output may change as a consequence of change in time preferences. Again, this is the keystone of Austrian macroeconomic theory. The underlying trade-off is between present and expected future consumption. The chain of causation proceeds from time preferences to the volume of savings, from savings to the level and structure of investment, and ultimately from investment to the pattern of employment.[5]

The analysis focuses on the consequences of a fall in time preferences, that is, a decrease in current consumption and a concomitant increase in expected future consumption. A decrease in the demand for consumption goods in the current period would be accompanied by an increase in the supply of loanable funds. In the absence of policy disturbances, and assuming no changes in liquidity and leisure preferences, this would result in a decrease in the rate of interest for loans and an increase in the quantity of loanable funds demanded. The lower rate of interest and greater availability of credit would provide an incentive to allocate more capital to the early stages of production. This is because the value of capital goods used in the early stages of production (fixed capital) is more sensitive to changes in the rate of interest than is the value of capital goods used in the later stages (circulating capital).

The reallocation of capital corresponds to an increase in demand for consumption goods in the future. A fall in the level of current consumption will release the resources needed for capital restructuring. The adjustment of the production process to the shift in time preferences will depend on the extent to which the entrepreneurs will recognize the profit opportunities and act upon them in the market process. We will return to this issue shortly.

Of course, there is no reason to expect that the adjustment process will run smoothly. The market process affords many opportunities for entrepreneurial error. However, there is no reason to believe that entrepreneurial error will be systematic, either. The market system does not necessarily result in either general gluts or shortages of resources. The historically observed "clusters" of entrepreneurial errors, that is, downturns in the business cycle, may be due to monetary policy exogenous to the market process. For example, a policy-induced expansion of credit and a concomitant drop in the rate of interest for loans may have the same initial effect as a fall in time preferences. Such a policy may be elected in a misguided attempt to stimulate economic growth through increased investment. However, this exercise of policy will ultimately fail because it does not release the resources required for a sustained restructuring of production.[6]

SUBJECTIVIST CONCEPTION OF BUSINESS CYCLES

Let us explore in some detail the consequences of a policy-induced expansion of credit, that is, a monetary disturbance.[7] How will such a policy affect the structure of capital goods? As will be argued below, the underlying theory is that of cyclical malinvestment, rather than overinvestment or underinvestment.

We have already seen that business cycles involve periods in which plans are persistently discoordinated. A mismatch of inputs to outputs results in unemployed resources. Plan discoordination entails entrepreneurial error, discussed in Chap. 2. Entrepreneurs who can reallocate resources to superior uses will earn profits, while those who fail to do so will suffer losses. As long as final output is scarce, there is no reason to expect that convertible resources will be persistently unemployed because cyclical variations generate profit opportunities.

We are interested here primarily in entrepreneurial decisions concerning the distribution of long-lived and short-lived capital goods, especially insofar as they affect decisions to utilize or abandon durable capital goods. The extent to which durable capital goods are convertible to other uses is of central importance in the study of economic fluctuations from the vantage point of building economics.

The analysis that follows is based on the distinction between *ex ante* and *ex post* entrepreneurial errors. On the one hand, *ex ante* errors involve errors in planning, that is, a misperception of future market conditions. As was argued in Chap. 2, the knowledge acquired during plan implementation changes nothing in this case, except that entrepreneurs become increasingly aware of their errors. Such errors may derive from wrong signals received by entrepreneurs, leading to false expectations. *Ex post* errors, on the other hand, involve errors in plan revision, that is, a failure to revise plans in accordance with actual market conditions. A theory of cyclical fluctuations must incorporate expectational errors. The theory expounded here involves expectational errors regarding money, that is, changes in the relationship between supply and demand in the money market.

Now, the interest rate mechanism coordinates saving and investment decisions.[8] This is illustrated in Fig. 4.1. The market interest rate is determined by the supply of and demand for loanable funds. At the "natural rate" (or the equilibrium rate), the plans of savers and investors are consistent with each other. An interest rate equal to the natural rate represents neutral bank policy, in which banks play the role of financial intermediaries between savers and investors. Under neutral policy, the supply of credit is determined by the supply of planned saving. At an interest rate below the natural rate, however, planned investment will exceed planned saving. We should note in passing that the con-

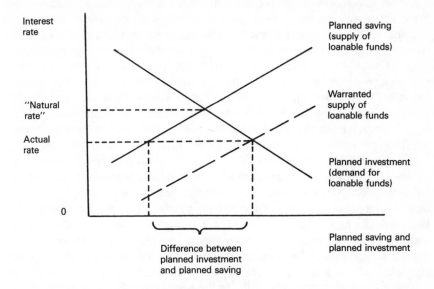

Figure 4.1 Neutral and nonneutral bank policy.

verse case, where the interest rate is above the natural rate, will not be discussed here, because it is not relevant to the argument. Now, a rate of interest below the natural rate can persist only if the difference is made up by the creation of bank credit. In such a case the bank policy is not neutral. Nonneutral bank policy leads to plan discoordination between savers and investors.

If a nonneutral bank policy is pursued for a period of time long enough to be perceived as a permanent change in business conditions, entrepreneurs will reshuffle their capital combinations to shift their output toward a more distant future. The share of durable production goods in investment projects will increase, ultimately leading to a decrease in the output of consumption goods. However, there will be no change in consumers' plans. By assumption, the interest rate has decreased due to the creation of bank credit, rather than in response to the shift in time preferences. Therefore, producers' plans and consumers' plans will be temporally out of alignment. As plan discoordination cannot persist for very long, some form of economic adjustment will be necessary.

The dynamics of expansion involve a transmission mechanism that goes from money, to incomes, to prices, and finally to output. First, business credit is expanded through the creation of bank money. Then, entrepreneurs spend the money on factor services, resulting in an increase in incomes. Finally, income recipients bid up the prices of consumption goods, thus affecting output. We will return to the dynamics of expansion after considering the impact of interest rates on capital structure.

Let us consider the impact of a fall in interest rates, resulting from a credit expansion policy, on the value of investment projects or existing capital combinations. The entire capital structure will ultimately be affected. Other things being equal, the decline in interest rates will raise the value of long-lived capital goods relative to the value of short-lived capital goods. The former yield consumption output in the more distant future than the latter. However, this is a partial effect only. The full impact of the fall in interest rates has three distinct types of effects: a discount rate effect, derived-demand effects, and cost effects.[9]

Now, returning to the dynamics of expansion, the conflict between *ex ante* returns in production goods industries, due to the policy-induced reduction in the interest rate, and *ex post* returns in consumption goods industries, due to rising incomes, will at some point result in a bidding war over resources between these industries. The prices of complementary goods needed for the completion of the new investment projects, or for the reshuffling of the existing capital combinations, will rise as a consequence of this competition for resources. This is especially true for the prices of complementary factors, such as labor and raw materials. Ultimately, the interest rates will increase, as the competition for loanable funds itself increases in the course of the cycle. Although the expansion process may at first proceed rapidly, in the later stages of a policy-induced expansion an increasing share of new investment projects involving long-lived capital goods will turn out to be unprofitable. The expansion will ultimately be checked, leaving behind a large number of unprofitable investment projects at various stages of completion.

This malinvestment cycle, engendered by a credit expansion policy, crucially depends on the complementarity of capital goods. As an investment project nears completion, the resources needed to complete it will have an increasing value. Its limit is the present value of the net proceeds the entrepreneurs expect from the entire investment project.[10]

FALSE EXPECTATIONS AND CAPITAL MALINVESTMENT

What effect will the current production of capital goods have on future demand for loanable funds? We will consider this question in the context of an economic upswing due to an unmaintainable drop in the interest rate. Again, special attention will be paid to the relationship between short-lived and long-lived capital goods.[11]

Most investment projects are undertaken in the expectation that further investment will take place at a later date. It is useful to think of investment projects as links in a chain that can be completed by a sequence of interrelated projects. These projects may be undertaken by different entrepreneurs. The entire chain can be characterized by the complementarity of individual capital goods that comprise it. Here, we are concerned primarily with diachronic complementarity, as distinct from synchronic complementarity at the focus of analysis thus far. The first link in an investment chain will be undertaken only in the expectation that a certain rate of return, that is, rate of interest, can be earned throughout the investment process.

Of course, the marginal investment projects that serve as starting points for entire investment chains will consist largely of fixed capital, while the subsequent investments will include an increasing share of circulating capital. Put differently, the proportion between long-lived and short-lived capital goods will tend to decline as the investment process unfolds.

Once the initial investment is undertaken, what are the consequences of an unexpected increase in the interest rate at which money can be borrowed? There is no necessary reason that the investment chain will be broken by any such unfavorable change in business conditions. The first step in the chain of investment projects narrows the margin within which the total profits expected on the entire chain may fall. However, the profitability of further investment projects needed to complete the chain may remain unaffected by the reduced profitability of the initial project. If the fixed capital initially created is specific to a particular purpose, and if it is difficult to convert it to alternative uses, it will be used even if the return barely covers the cost of using it (excluding interest and amortization). As a consequence, the entrepreneurs contemplating investment projects that would complete the chain may be enticed to undertake them precisely because the products of the earlier stages of production may be lower in cost than initially expected.

It follows that the demand for capital, that is, for loanable funds available for investment purposes, will depend largely on the proportion of capital goods already in existence relative to those capital goods needed to complete various investment chains. The impact of an increase in the rate at which capital can be borrowed will directly affect only a small fraction of total investments—those that represent the initial stages of new investment chains. An unfavorable change in business conditions will undoubtedly have a dampening effect on such marginal projects. For the projects needed to complete the existing investment chains, the demand for capital will tend to be inelastic with respect to changes in the interest rate.

Where the process will end depends on how rapidly the demand for capital drives up the interest rate, as well as on how much of a given capital structure has already been completed. As some investment chains near completion, the demand for capital to acquire the complementary capital goods will tend to subside. However, many such chains will not be completed due to the scarcity of complementary capital. Both diachronic and synchronic complementarity are of interest in this regard. The profitability of the capital goods created at the early stages of the process will progressively decrease as the interest rate rises. Malinvestment in these capital goods will become increasingly apparent, and will contribute to the ultimate collapse of the whole process.

Let us briefly return to the relationship between long-lived and short-lived capital goods. As we have already seen, long-lived capital goods will be more affected by increases in the interest rate than will short-lived capital goods. As the investment process unfolds, each completed investment chain will therefore contain an ever-decreasing proportion of the former relative to the latter. Furthermore, unfavorable business conditions will lead prudent entrepreneurs to revise their production plans to complete the investment chains earlier than originally anticipated. This will lead to an unanticipated shift from long-lived capital goods—which will be converted to new uses whenever possible—to short-lived capital goods needed for the completion of investment chains. This amounts to a new round of capital restructuring. As we have seen in Chap. 2, this applies to buildings at least in the sense that they are adaptable and flexible at a *given* location, although they are locationally inconvertible.

Under what conditions will the demand for capital required to complete a given capital structure force the rate of interest to a level higher than may be permanently maintained? The process described here arises from any change in business conditions leading entrepreneurs to expect a lower interest rate or a larger supply of loanable funds than may exist when investment projects become available for utilization. These false expectations will ultimately result in a greater increase in the interest rate than would have occurred had there been no initial expectations of such a low rate. Although several causes may create such a situation—such as an unexpected drop in the rate of saving, or an unforeseen innovation demanding new capital—the most common cause of widespread entrepreneurial errors is an unmaintainable increase in the supply of loanable funds through credit expansion.[12]

An artificial upswing of the type described here will be checked by the scarcity of loanable funds needed for the completion of various investment chains. At first glance, this involves a paradox: the abundance of unused capital goods is in fact due to the scarcity of capital required to complete investment chains started under erroneous expectation of a permanent change in the availability of loanable funds, that is, under the erroneous assumption that the change represents a shift in consumers' time preferences.[13]

We have already discussed a policy-induced expansion of credit and the adverse effect it is likely to have in terms of economic fluctuations. Here we have explored the mechanism of the malinvestment cycle in some detail. It is clear that capital malinvestment, based on false expectations of entrepreneurs, will have an especially adverse effect on the most durable capital goods, such as buildings, to the extent that they cannot be converted to new uses. This may result either from the inherent inconvertibility of some capital goods, or from the unavailability of resources required for conversion of those capital goods that are "technically" convertible.

AFTERMATH OF THE MALINVESTMENT CYCLE

A process of contraction will follow an abortive expansion. How will contraction look? Faced with the scarcity of complementary goods and factors, entrepreneurs will be forced to abandon their plans to complete many investment projects. The prohibitively high prices of labor and raw materials will lead to their release from the production of capital goods. This will precipitate an initial round of unemployment. The resulting unemployment will eventually spill over into the production of consumption goods. Cumulative contraction may develop if monetary policy does not respond quickly enough [O'Driscoll and Rizzo, 1985: 210].[14]

Now, the investment ceiling leading to contraction consists of two subceilings, one for raw materials and another for fixed capital goods. The former will be hit first, followed shortly by the latter. At first, the fixed capital goods subceiling will be revealed in the form of delayed deliveries. Delivery delays can only postpone the appearance of excess capacity in industries producing fixed capital goods. Higher prices will follow. As we have already seen, the scarcity of complementary factors will ultimately lead to the emergence of excess capacity throughout the economy [Lachmann, 1978a: 107].

At first, excess capacity translates directly into a drop in the rate of capital utilization. This is because the volume of fixed capital, consisting mainly of plant and equipment, cannot adjust readily to output fluctuations. As Kalecki [1971: 135–36] argues, "the volume of fixed capital fluctuates relatively little in the course of the cycle so that fluctuations in output reflect mainly changes in the degree of utilization of equipment." If excess capacity problems are severe, the process of adjustment will have to proceed beyond reductions in the rate of capital utilization.

Capital malinvestment will ultimately require adjustments in the capital structure. However, these adjustments cannot be made without additional investment expenditures. Due to the heterogeneity of capital, malinvestment in one period cannot be removed in the next period through disinvestment. The process of adjustment will require considerable time for plan revision and for the accumulation of resources needed for capital regrouping. Meanwhile, the existing capital must be used in ways not originally planned.[15]

The period of adjustment will affect the long-lived capital goods most. A good deal of the malinvested capital value will be lost before its planned depreciation period can be completed. The entrepreneurs owning these capital goods will have to find new uses for them. In some of these new uses, long-lived capital goods will be more profitable than originally envisaged, while in others their profitability will fall short of expectations. The key point here is that durable capital goods are especially susceptible to losses in value due to unexpected changes in business conditions. In fact, it is often impossible to make meaningful plans over the entire life cycles of the most durable capital goods, such as buildings, because change cannot always be foreseen. According to Lachmann [1978a: 37–38]:

> The loss in value of course reflects the fact that capital instruments, particularly those that are durable, have to be used in ways other than those for which they were

designed. In these new uses the instruments may be either more or less profitable than in their designed uses. In the former case there will be gain, in the latter a loss of value, i.e., their market value will differ from their cost of production. The cause of the phenomenon is unexpected change. Hence, durable capital goods are more likely to be affected than those more short-lived. In the case of buildings our phenomenon often occurs for the simple reason that they last for longer periods than could possibly enter any plan-maker's "horizon." Often, as we stroll in the streets of an ancient town, the merchants' palaces turned into hotels, the former stables now garages, and the old warehouses which have become modern workshops, remind us of the impossibility of planning for the remote future.

Existing capital goods tie the present to the past. They introduce something akin to inertia into the economic process. When their conversion potential is limited, they restrict our field of action. In Mises' [1966: 506] words:

> Capital goods are a conservative element. They force us to adjust our actions to conditions brought about by our own conduct in earlier days and by the thinking, choosing and acting of bygone generations.

Capital goods that cannot be turned to new and profitable uses will be either operated but not maintained or abandoned altogether. The owners of such capital goods will find it unprofitable to maintain them. In fact, this may be the best remedy for capital malinvestment.[16] During contraction, capital maintenance and replacement activities will generally be neglected, thus providing an essential precondition for the eventual expansion of investment activity sometime in the future. This issue will be taken up shortly.

MALINVESTMENT AND LEARNING

The malinvestment cycle presented here should not be construed as a process that forever remains unintelligible to economic agents. We cannot believe that these agents will respond to successive cycles in identical fashion. People do learn from their experience, and the cycle cannot be reproduced time after time. In general, as Shackle [1969: 57] argues,

> The one-way traffic of human history allows no repetition of the kinds of experiments which change man's ideas. They are self-destructive experiments.

Moreover, it is likely that the monetary theory of business cycles has also contributed to a less sanguine reaction to credit expansion [Mises, 1966: 797, and Lachmann, 1986: 30]. The naive optimism associated with credit expansion in the past has given way to a certain skepticism among entrepreneurs. As a result, the length of the cycle has been considerably reduced. Although the alternation of boom and slump has remained, their amplitude is less pronounced and their succession is more frequent. For these reasons the business cycles tend to be less predictable, as well.

The most important consequence of entrepreneurs' skepticism concerning credit expansion is that the amount of malinvestment is smaller, as the boom comes to an earlier end. The following depression is therefore milder, too [Mises, 1966: 798]. Of course, this is of utmost importance in the case of long-lived capital goods. First, major capital projects will be undertaken only when economic agents are convinced that the boom is indeed likely to last long enough for the completion of their projects. Second, especially large projects will be phased to straddle two or more booms of shorter duration. This has become a rather common feature of speculative real estate development projects, for example. We will return to this subject in the last section, where we will focus on changes in building technology conducive to rapid capital restructuring.

THE BUILDING CYCLE

How will the expansion and contraction processes associated with a strong boom look to building owners and building professionals—real estate developers, architectural and engineering designers, and contractors? Let us reconsider these processes from a different angle, that of the building cycle. The sequence of events presented here corresponds to the sequence of events in the malinvestment cycle. We have already offered the theoretical underpinnings of the story that follows in the preceding sections. We should bear in mind that not all building cycles will be identical in terms of their pattern and intensity, as the underlying business cycles are by no means identical, either.

In the early stages of credit expansion, business conditions will temporarily favor long-lived capital goods. The rate of utilization of fixed capital will increase. The lower interest rate and the greater availability of loanable funds will in due course entice building clients to start considering new investment projects. New investment chains will require new space, and clients will form building committees to consider a variety of building ventures. The land market will tighten after land held in reserve is committed to new projects. Real-estate developers will consider large speculative projects such as office buildings, industrial parks, and shopping centers. They will also enter joint ventures with clients looking for ways to spread the risk inherent in their building plans.

Buildings suffering from disrepair because of low levels of maintenance and replacement expenditures over a protracted period following the previous boom will be considered for renovation in the context of new investment projects. Space held in reserve will be allocated to new uses. The efficiency of space utilization will be improved by gradually eliminating space "waste." Initially, design and construction projects will focus on the existing properties in order to convert them to new functions, but the expansion will soon require new land and new constructed facilities.

Design firms of all sizes will receive an increasing number of projects. They will, in turn, hire a variety of building consultants to advise them on specialized equipment, building systems, and construction techniques. Under pressure of an increasing number of projects, design firms will hire additional people to complete design teams of all sizes, but many aspects of new projects will be subcontracted to specialized consulting firms.

As design projects near completion, designers will advise their clients on choosing contractors to undertake the construction jobs. At first, building clients will readily acquire the necessary financing and then proceed to arrange the building contracts. As contractors receive new orders, they will need to hire additional management personnel, as well as buy some new construction equipment; however, most of the required capital equipment will initially be rented or leased. Although the contractors will hire some additional construction workers of all trades, they will rely mostly on a large number of specialized subcontractors with whom they have established good working relations in the past. In turn, the subcontractors will be hiring construction workers to perform the work required.

Wholesalers and retailers of building materials will be receiving new orders. At first, they will run down their inventories, but as soon as they realize that the market has tightened for a long haul, they will replenish their stocks. The same is true of the manufacturers of building materials, who will ultimately increase the rate of production and adjust their inventories to the new level of construction activity.

New land will be developed, and an increasing proportion of derelict property will be cleared and made available for new building projects. Construction sites will spread, and construction crews will work day and night. As new buildings sprout at an increasing rate, a fresh sense of optimism will gradually spread through all the building professions. People who had been unable to find jobs will be fully engaged by now, with a prospect of permanent employment in growing firms.

Throughout the early stages of expansion, all the economic agents involved will experience increasing pressure to expand their own facilities. The increasing scale and rate of their operations will require new space. This will cause a secondary wave of building projects to spread through the economy.

At some point, however, some of the economic agents involved in the ever more intensive building activity will start experiencing difficulties with the expansion of their operations. They will learn that a growing number of people hired do not perform up to expectations, that wages and salaries are steadily increasing, that organizational problems connected with larger firms or their divisions require new and untried management arrangements, and that many of the services they have been accustomed to have become increasingly scarce. In consequence, they will still accept new orders, but they will let the backlog pile up. The order books will grow for a while before these economic agents will consider raising the prices of their services. One of the justifications for price increases will be that their costs are growing, too.[17] The cumulative process of price increases will have started, most likely led by wages and prices of building materials.

Meanwhile, building clients will begin experiencing problems with financing their investment projects. The interest rate will climb under pressure of many competing projects, but loanable funds will still be available. The combination of rising interest rates and spreading price increases will become increasingly worrisome. Some of the building projects nearing completion will turn out to be somewhat less profitable than their builders had hoped. Others will prove to involve outright losses. This will dampen the initial enthusiasm for new ventures involving considerable investment in fixed capital.

An increasing number of design projects completed by the design firms will be waiting for the initiation of construction. The designers will be busy administering the ongoing building projects, as the construction period generally stretches over a number of years. The backlog of new orders for design projects will at some point start dwindling. Plans for further expansion of design firms will be seriously reconsidered, and many of them will be postponed.

The contractors, whose resources will have been spread fairly thin by the expansion, will experience increasing management problems. The coordination of an increasing number of ongoing projects will become more difficult as the process unfolds. Some contractors may run into financial trouble, as many overlapping construction projects strain their ability to obtain construction loans for interim financing of these projects at reasonable interest rates. This will lead to cash-flow difficulties. Bankruptcies will spread, although the construction firms going under would have made handsome profits had they been able to finish the ongoing projects. Moreover, some clients will be unable to pay for construction services in accordance with established schedules because of their own financial difficulties, and that will put an additional strain on the financial management of construction firms.

It is likely that contractors will experience the greatest need for capacity expansion precisely when the economic cycle has already passed its peak. Many construction projects will continue even after the boom has been checked, because the construction period for significant projects tends to be rather lengthy. The backlog of construction projects will stop increasing well before the peak of the cycle, as new building orders cluster in the early stages of expansion, but an increasing number of projects will be taken off the order books as the financial situation worsens. Some contracts will be cancelled even before the construction has started, which will reduce the pressure but also signal bad times ahead.

As we have argued repeatedly in the preceding sections, increases in the interest rate will affect long-lived capital goods, such as buildings, to a greater extent than short-lived capital goods. In fact, some construction projects will be interrupted in mid-course, as clients reconsider their plans and find new investment projects unprofitable. Nevertheless, many construction projects will proceed even after the collapse of the expansion process. Contractors will have to complete these projects while refraining from enlarging their operations beyond the minimum necessary for project completion. The conclusion of the business cycle will already be apparent around them, and they will have to prepare for contraction at a time when they are extremely busy with ongoing projects at various stages of completion.

The contraction process will expose the malinvestment of building and other resources. Many buildings, constructed as part of investment chains long abandoned as unprofitable, will require redesign and reconstruction even before they have entered the production process. Building clients will have difficulties scraping together the funds required for these activities. Nevertheless, the aftermath of the malinvestment cycle will involve redesign and reconstruction on a significant scale. In fact, it is likely that such activities will start well before the expansion has collapsed, as an increasing number of building clients will have realized that their initial plans could not be implemented as originally conceived.

Depending on the strength of the boom that initiated building projects of all kinds and sizes, the contraction will be followed by a more or less lengthy period over which maintenance and replacement activities will be kept at a minimum. Existing buildings will gradually deteriorate to the point at which their redesign and reconstruction may help generate a new expansion.[18] During the period of relatively low economic activity, there will nevertheless be some building activity, but it will focus on conversion rather than new building ventures.

During the protracted downturn of the business cycle, the firms providing all types of professional services in the building field will be reduced to the minimum size required for smooth operation. Contractors will release specialized subcontractors, with whom they had established nearly permanent relations during the expansion.[19] To a lesser degree, this holds for designers, as well. The competition for building projects will reduce the prices of these services to some extent, and this will also have an impact on salaries and wages. The same is true of building material prices because wholesalers and retailers, as well as manufacturers of building materials, will find it difficult to place their products in shrinking markets. However, it is very likely that the main adjustment mechanism will be through the quantities of goods and services offered, rather than their prices, which are notoriously sticky on the downward side.

CAPITAL MAINTENANCE, OBSOLESCENCE, AND TECHNICAL CHANGE

We have seen that the duration of an economic downturn depends on the durability of capital goods, which in turn depends on capital maintenance. If the income stream generated by a capital good is not high enough, the capital good will be undermaintained. Its useful life will thus be foreshortened. In addition, the capital good might not be replaced in full at the end of its life cycle. Let us explore this connection in greater detail, with a special emphasis on technical innovation in general, and innovations in building technology in particular.

In economic terms, the problem of capital maintenance translates into the problem of maintaining a permanent income stream from capital. This subject was already broached in Chap. 3. Of course, this income stream does not necessarily have to be constant; it may have any desired time-profile. The reason for desiring that capital be maintained in a particular way is that it may be unintentionally depleted if an adequate part of the income stream is not used for its maintenance because of an error in reckoning net income. One of the crucial roles of capital accounting is to make sure that involuntary infringements upon future income do not occur [Hayek, 1975: 94].[20]

Two forms of regularly occurring destruction of the existing capital values may be distinguished: wear and tear, and obsolescence. Wear and tear depends on the rate of capital utilization, and requires that an amount proportional to it be put aside into a sinking fund that will accumulate over the life of a particular capital good, in order to secure its eventual replacement. In addition, a part of the income will have to be put aside for current maintenance, assuring the smooth operation of the capital good. Obsolescence of real capital occurs whenever its usefulness diminishes faster than it deteriorates physically. In such a case the amount to be put aside should be proportional to the decrease in the value of the investment, over and above wear and tear.

It is interesting to note that obsolescence would be a problem even in the theoretically useful case of perfect entrepreneurial foresight, because in many cases fixed capital must be made more durable than necessary for purely technical reasons. Even if it is needed only for a short period of time, some of its performance requirements may include a combination of physical properties—such as strength or hardness—that give it a life cycle exceeding the requirements. As Hayek [1975: 105–6] argues:

> There can be no doubt that many investments in actual life are made with complete awareness of the fact that the period, during which the instrument concerned will be useful, will be much shorter than its possible physical duration. In the case of most very durable constructions like the permanent way of a railroad, the prospective "economic life" ought to be regarded as considerably shorter than the possible "physical life." [. . .] It is impossible to adjust the durability of a machine to the short period during which it may be needed, and in many other cases the strength needed from a construction while it is used necessitates it being made in a form which will last much longer than the period during which it is needed.

Much technological innovation concerns finding a better match between the economic and physical lives of durable capital goods. We will return to this subject shortly, as it is undoubtedly of special importance in the case of constructed facilities in general, and buildings in particular.

The case of imperfect entrepreneurial foresight is of course of much greater practical importance. A capital good may be given greater durability than will turn out to have been needed. Put differently, had future economic conditions been predicted correctly, the capital good rendered obsolete by subsequent events might have been designed to be less durable [Hayek, 1975: 108].

As we have already seen, the malinvestment cycle promoted by a monetary disturbance offers an example of unexpected change leading to widespread obsolescence of capital goods, especially long-lived capital goods. Among other kinds of unexpected change that may affect the value of a capital good, technical innovation is perhaps the most interesting [Hayek, 1975: 113–15]. An invention may either increase or decrease the value of existing capital. This will depend on whether the new capital goods required by the invention and the existing capital goods are complements or substitutes of one another.

The capital gains that will arise in the former case will tend to be temporary, and will accrue to the owners of these capital goods over a period of time needed for capital restructuring promoted by the invention. In the latter case, capital losses will tend to be permanent. Generally, a significant invention is likely to result in an increase in demand for capital, which will in turn cause the interest rate to rise. This will affect the profitability of the existing fixed capital. Capital losses will be most pronounced in the case of durable capital goods, because they tend to be most sensitive to changes in the interest rate.

Of course, capital losses resulting from the invention will reduce the profitability of some capital goods, and render others altogether unprofitable. In the latter case, the capital goods will be abandoned, while in the former their maintenance will be threatened. The income streams from capital goods anticipated by the entrepreneurs will be reduced, thus leading to a reduction in capital maintenance, that is, to more rapid physical deterioration of these capital goods. Therefore, an invention will affect a part of the existing capital through both the reduction of funds available for capital maintenance and the reduction of its value due to obsolescence.

Now, it is usually assumed that there is no great difficulty in distinguishing between the normal processes of maintaining and replacing the existing capital, on the one hand, and net addition to it, on the other. In other words, it is assumed that it can be easily determined whether capital increases, decreases, or remains constant. However, in a changing world, characterized by imperfect foresight, such determinations will depend on the subjective judgment of individual entrepreneurs. The meaning of "maintaining capital intact" depends on how well an individual entrepreneur foresees the future [Hayek, 1975: 116].

BUILDING TECHNOLOGY AND CAPITAL RESTRUCTURING

The uncertainty associated with economic fluctuations will provide an incentive for entrepreneurs to adopt general strategies that may offset the threat of unexpected change in business conditions. We have already encountered this issue in the discussion of malinvestment and learning. In this context, it is reasonable to assume that entrepreneurs will generally prefer short-lived to long-lived capital goods in their specifications of capital combinations, if such a choice is available to them. Furthermore, they will generally prefer the least durable among the long-lived capital goods that perform the same function, subject to the condition that the expected life cycles of these capital goods extend beyond the time-horizon they consider relevant for planning purposes. This condition will lead the entrepreneurs to put a special premium on the predictability of the life cycles of capital goods or their components.

Applying this reasoning to investment in "plant and equipment," two inter-related secular tendencies may be conjectured. First, we may expect a tendency for investment in production equipment to increase at a faster rate than investment in plant.[21] Second, regarding the investment in plant itself, we may conjecture that there will be a tendency toward an increasing share of mechanical, electrical, and electronic equipment, that is, a decreasing share of "bricks and mortar" in building investment. Moreover, it is reasonable to expect that the life cycles of plant and equipment will tend to converge. That is, there will be an increase in the number of innovations in building technology that simultaneously increase the reliability and reduce the life expectancies of the most durable among building components and systems, such as structural systems, exterior cladding, and roofing—components of the so-called building shell. As already mentioned, this would make the predictability of the life cycles of building components and systems increasingly important.

Regarding the construction process, that is, the assembly of building components on site, the above tendencies translate into a shorter construction period. Much innovation in building technology, in fact, concerns this aspect of the building process. Construction period shortening reduces the time-interval over which a project is vulnerable to sudden changes in economic conditions, as we have seen in Chap. 3.

Another general strategy that is of interest here concerns the rate of capital utilization, measured by the number of operating hours per week, for example. It is worth noting that "[t]he number of hours per week a business establishment is ordinarily open and operating is an aspect of the firm's investment decision" [Foss, 1984: 5]. The rate of capital utilization is therefore an aspect of the production plan. Foss [1963 and 1984] provides considerable empirical evidence of a secular tendency for the rate of capital utilization to increase.[22] It is plausible to hypothesize that capital durability and the rate of capital utilization are inversely related, other things remaining equal. A capital good with a longer workweek will tend to have a shorter life because of an increase in wear and tear.[23] Of course, this effect can be offset by a corresponding increase in maintenance and replacement expenditures, but in that case an increasing threat of technical obsolescence will arise. Therefore, we may conjecture that the increasing rate of capital utilization is partly due to the desire to depreciate the capital assets faster in view of pervasive economic change. Again, this will have a disproportionate effect on long-lived relative to short-lived capital goods, because the durability of the latter will dictate the useful life of the former. In terms of building technology, the conclusions reached above apply here, as well.

In a world characterized by unexpected change, the above propensities undoubtedly have a rational basis. Reduced durability of long-lived capital assets may contribute to a smoother process of capital restructuring, and thus to a less painful economic recovery from recessions and depressions, that is, from malinvestment cycles of different degrees of severity. Although flexibility and adaptability undoubtedly remain important desiderata for buildings, as we will continue to argue in Chap. 5, these characteristics refer only to buildings themselves; they imply greater rigidity in the large-scale environment, because buildings are characterized by locational immobility. Tendencies toward a reduction in building durability do not suffer from this defect; however, their effect on the stability of the built environment may have far-reaching consequences outside the economic domain. Some degree of perceptual stability of the built environment may be desirable in itself.[24]

CONCLUSION

Buildings, as well as other long-lived capital goods, are especially sensitive to economic fluctuations. Shifts in time preferences in the absence of policy disturbances should have an appreciable effect on building activity, be they perceived as favorable or unfavorable by those engaged in the building industry. However, the consequences of policy-induced disturbances are, in fact, particularly deleterious in the building realm.

The building owners will have to reshuffle their capital combinations, including real property, in the aftermath of a malinvestment cycle. The process of capital restructuring following upon an abortive upswing may result in a great deal of rebuilding, but a significant portion of building resources will also be wasted. The market process ensures that some buildings erected on the basis of false expectations will find unexpected buyers or users, as was argued in Chap. 2, but this argument cannot be extended to all building malinvestment that may take place under conditions of policy-induced credit expansion.

The threat of building malinvestment motivates the building owners to adopt general approaches that offset the underlying uncertainty. The building fabric, configuration, and structure will be modified to reflect these concerns. This will affect the development of building technology and ultimately building design, as we have seen in Chap. 3.

In an attempt to control their fate, those involved in different phases or facets of the building process will also engage in various types of economic evaluation of their projects. For this purpose, they will seek analytical frameworks sensitive to time, that is, to structural or qualitative changes in the relationships between the economic variables considered. In Chap. 5, we will explore some methodological approaches that satisfy this requirement. In particular, we will discuss some tools and techniques that are especially useful for the representation of building processes, both from the vantage point of an individual building and the entire real property holdings at the disposal of an owner. Special attention will be given to the research tasks facing building economics as a field, both from the vantage point of methodological and substantive issues.

NOTES

1. It should be borne in mind that this distinction between macroeconomic and microeconomic domains is of heuristic value only. The Austrian school rejects any dichotomization of these two domains, because changes in macroeconomic aggregates cannot be understood independently of individual action. The concept of "national economy," and the associated macroeconomic aggregates, would be permissible in economics only if the nation in its totality were really the economic subject [Menger, 1985: 193]. This is decidedly not the case, however.

2. This overview is based on Garrison [1985].

3. It is interesting to note in this connection that Ricardo [1951: 31n, 38, 150] distinguishes between circulating and fixed capital on the basis of its durability. As Ricardo [1951: 38] argues, "[i]n proportion as fixed capital is less durable, it approaches to the nature of circulating capital." Of course, this is the reason why circulating and fixed capital cannot be definitely demarcated, "for there are almost infinite degrees of durability of capital" [Ricardo, 1951: 150]. (See note 11 in Chap. 2.)

4. However, Garrison [1985: 174] warns:

   The relationship between the demand for money and the structure of production identified by Kessel and Alchian deserves attention, even though the effect of the life of capital goods on the demand for transaction balances has not been clearly established. The existence of long-lived capital goods that change hands a number of times blurs the distinction on which the relationship rests. Conversely, the choice within a single firm between the production and employment of one long-lived capital good and the production and employment of a sequence of short-lived capital goods is unrelated to transaction balances. It can be noted in passing, however, that to the extent that long-lived capital goods are associated with early stages of production (physical plant) and short-lived capital goods are associated with the later stages (working capital), many of the implications of the Kessel-Alchian analysis are consistent with those of the Austrian analysis.

5. Note that Garrison [1985: 177] carefully points out the limitations of the analysis that follows:

   The present concern [. . .] is how the various markets *would have to work* if a change in time preferences on the part of consumers is to be successfully translated into a corresponding change in the pattern of production activities. The issue of whether the markets *do* in fact work in this way [. . .] is an issue that turns on our understanding of entrepreneurship. The following discussion will assume that entrepreneurs are alive and well and that they tend to discover and act upon opportunities for making profit.

6. For this reason the Austrian theory of capital is intimately related to both monetary and business cycle theories.

7. This overview is based on O'Driscoll and Rizzo [1985: 198–211]. Figure 4.1 is also based on O'Driscoll and Rizzo [1985: 203]. Because of the technical complexity of the argument, only a highly simplified version of it will be presented here.

8. According to O'Driscoll and Rizzo [1985: 202], this position is associated with the Thornton-Wicksell monetary tradition. Note that Wicksell was influenced by Böhm-Bawerk, who was in turn influenced by Menger. The authors state that they follow the Austrian development of that tradition, originally enunciated by Mises and Hayek. For a historical discussion of this theoretical development, see Schumpeter [1954: 720–25 and 1118–21], for example.

9. For further details, see O'Driscoll and Rizzo [1985: 205], who elaborate the relative roles of these three effects, which may offset each other. Briefly, the discount rate effect follows from the negative elasticity of present value with respect to the interest rate; the derived demand effects stem from the changes in relative profitability of long-lived and short-lived capital goods; and cost effects are due to the fact that various bottlenecks encountered during an expansion drive up the prices of complementary factors, especially labor and raw materials. It should be noted, however, that throughout this analysis the authors are not concerned with the demand for capital as such, but rather with changes in the pattern of investment flows [O'Driscoll and Rizzo, 1985: 206].

10. As O'Driscoll and Rizzo [1985: 209] write:

> Indeed, it was precisely this consideration that led Hayek [1975: 73–82] to one of his most important analytical contributions to business cycle theory. The capital complementarity effect helps explain the pro-cyclical behavior of interest rates, *apart from any Fisher effects*. A prolonged cycle of capital investment is likely to increase the expected returns from additional borrowing and investment. Past investment raises the demand for current investable funds, driving up real market interest rates. Toward the end of a cycle, the real short-run interest rate, which would clear the market, will be higher—perhaps significantly higher—than the long-run equilibrium rate.

11. This overview is based on one of Hayek's [1975: 73–82] seminal papers. This paper, entitled "Investment that Raises the Demand for Capital," was reprinted from *The Review of Economic Statistics*, vol. XIX, no. 4, November 1937.

12. According to Hayek [1975: 80–81]:

> It is not so much the quantity of current investment but the direction it takes—the *type* of capital goods being produced—which determines the amount of future in-

vestment required if the current investments are to be successfully incorporated in the structure of production. But it is the amount of investment made possible by the current supply of funds which determines expectations about the future rate of investment and thereby the form that the current investment will be given. [. . .] An increase in the rate of investment, or the quantity of capital goods, may have the effect of raising rather than lowering the rate of interest, if this increase has given rise to the expectation of a greater future supply of investible funds than is actually forthcoming.

Hayek's emphasis on differentiating the types of capital goods is particularly important here, because it illuminates the relationship between long-lived and short-lived capital goods. An unmaintainable drop in the interest rate will temporarily favor long-lived capital goods. However, the subsequent unexpected rise in the interest rate will render a large portion of these capital goods unprofitable, especially if the investment chain of which they are part cannot be completed. This is why we have emphasized malinvestment cycles and the problem of malinvested capital, which is most severe in the case of those long-lived capital goods that are highly specific and inconvertible. This applies to buildings, which are generally convertible to new uses, to the extent that malinvestment is associated with their location.

13. As Hayek [1975: 149] writes:

This phenomenon of a scarcity of capital making it impossible to use the existing capital equipment appears to me the central point of the true explanation of crises; and at the same time it is no doubt the one that rouses most objections and appears most improbable to the lay mind. That a scarcity of capital should lead to the existing capital goods remaining partly unused, that the abundance of capital goods should be a symptom of a shortage of capital, and that the cause of this should be not an insufficient but an excessive demand for consumers' goods, is apparently more than a theoretically untrained mind is readily persuaded to accept.

14. According to O'Driscoll and Rizzo [1985: 210–11], most of the decline will appear to be Keynesian (for an elaboration of Keynes's business cycle theory, see Keynes [1964: 313–31], for example). They point out that Keynesian analysis begins in the last stages in the expansion, when the marginal efficiency of investment declines. However, this decline is predicated upon the assumption that capital goods are homogeneous, that is, that they are substitutes for each other. The Austrian theory of economic fluctuations is based on the assumption that capital goods are heterogeneous. Their complementarity is an essential ingredient of the Austrian view, as we have seen throughout this book.

15. As Lachmann [1978a: 118] writes:

The essence of the matter is that investment decisions are not merely irreversible in time, so that excessive investment in period 1 as a rule cannot be offset by disinvestment in period 2, but that they are also irrevocable in kind. [. . .] If all capital were homogeneous there would be no sub-ceilings and the advance would not be halted until all resources had become equally scarce [. . .]. As it is, capital is heterogeneous, and the first sub-ceiling reached will necessitate not merely the revision of plans for the construction of new capital, but also the revision of plans for the use of existing capital.

16. As Lachmann [1978a: 126] argues:

> The best remedy for the excess capacity mentioned is to make it unprofitable for the owners of such resources to maintain them. Except in the case where there is excess capacity everywhere, the case in the contemplation of which the Keynesians specialize, existing capital combinations must be broken up and their fragments removed to wherever they are still useful.

17. The differentiation between the quantity and price signals, and an emphasis on the quantity signals in the market process, is emphasized by Kaldor [1985], for example. Quantity signals involve a change in the amount of stock carried, or a change in the size of the producer's order book. As Kaldor [1985: 24] argues:

> Stocks (or inventories) are carried by all those producers who make standard-type articles, where the buyer (whether an ultimate buyer, or an intermediary, a wholesale or a retail merchant) is accustomed to obtaining delivery within a short period, which is only possible by getting the particular article "off the shelf" of the manufacturer. Order books mainly relate to custom-built articles, whether ships, specially designed houses, or suits made by bespoke tailors, and where the buyer is expected to wait much longer for delivery after giving his orders, but if possible not too long, so that manufacturers of these latter categories of goods carry input stocks to save time and possible uncertainty of obtaining the inputs necessary for fabrication.

Kaldor [1985: 25] argues that "in the actual adjustment of supply and demand, prices play only a very subordinate role, if any," while "[q]uantity signals are invariably prompt."

18. According to Keynes [1964: 317]:

> The explanation of the *time-element* in the trade cycle, of the fact that an interval of time of a particular order of magnitude must usually elapse before recovery begins, is to be sought in the influences which govern the recovery of the marginal efficiency of capital. There are reasons, given firstly by the length of life of durable assets in relation to the normal rate of growth in a given epoch, and secondly by the carrying-costs of surplus stocks, why the duration of the downward movement should have an order of magnitude which is not fortuitous, which does not fluctuate between, say, one year this time and ten years next time, but which shows some regularity of habit between, let us say, three and five years.

Keynes [1964: 318] thus concludes that "the interval of time, which will have to elapse before the shortage of capital through use, decay and obsolescence causes a sufficiently obvious scarcity to increase the marginal efficiency, may be a somewhat stable function of the average durability of capital in a given epoch." The standard time-interval will change if the characteristics of the epoch shift.

19. According to Eccles [1981], the general contractor and specialized subcontractors form a stable relationship when conditions permit. This organizational form, called the "quasifirm," is analogous to the "inside contracting system," discussed by Williamson [1975 and 1979]. However, it should be pointed out that the quasifirm's size will vary with the building cycle, increasing during the boom, and decreasing during the slump. This variability is perhaps one of the main reasons for the existence of this flexible organizational form in the building industry.

20. See note 31 in Chap. 2. It should also be noted that this discussion concerns *ex ante* preservation of capital value. *Ex post* preservation of this value ultimately depends on the satisfaction of the wants of the consumers, as has been argued in Chap. 2. In Mises' [1978: 110–11] words:

> The owner of producers' goods is forced to employ them for the best possible satisfaction of the wants of the consumers. He forfeits his property if other people eclipse him by better serving the consumers. In the market economy property is acquired and preserved by serving the public and is lost when the public becomes dissatisfied with the way in which it is served.

21. This is in agreement with Maddison's [1987] analysis of longitudinal data for six advanced capitalist economies, as well as with conclusions reached by Ventre [1982], using the U.S. data.

22. On the rate of capital utilization, see also Kuznets [1961], Denison [1980], Betancourt and Clague [1981], and Winston [1982], for example.

23. This is in agreement with Bischoff and Kokkelenberg's [1987] analysis of the U.S. data.

24. On this issue, see Lynch [1972: 107–13], who explores a wide range of approaches to achieving the adaptability of the built environment while maintaining its coherence.

5

Toward a
Research Program

INTRODUCTION

In this chapter I will outline several research tasks that face building economists as professionals, and building economics as an emerging field. Two caveats are in order, however. First, this chapter offers a very personal view of the prospects ahead of us. Several alternative development paths for building economics will be introduced and discussed in considerable detail. I have already pursued some of these paths in my own work, or together with my colleagues at MIT. Second, this chapter focuses on a few issues that I consider to be of central importance, rather than on the entire field. The prospects outlined here are not meant to be exhaustive, because it is still too early for a panoramic view of building economics.

We will discuss both methodological and substantive issues, in that order, from the vantage point of an owner of buildings and land engaged in the production of goods and services. The importance of embedding economic theory in software tools will be emphasized. In a sense, some emerging varieties of these tools offer the possibility of "live" economic knowledge that combines economic theory with the subjective economic experience of individual users. These tools might thus serve as repositories of the users' learning about changing economic conditions. We will assume that the reader is familiar with basic notions of expert systems and decision-support systems, the first fruits of research in the so-called artificial intelligence that have reached the marketplace.[1]

The role of time in the study of the economic aspect of building processes deserves special attention, as has been argued repeatedly in the preceding chapters. Building decisions often involve qualitative choices concerning time, such as whether to build sooner or later. As people engaged in the building process tend to think about time in qualitative terms, building economics should be able to offer them useful advice in this regard. To paraphrase Hicks [1984: 280], it is because I want to make building economics more human that I want to make it more time-conscious. I will discuss some pitfalls of quantitative analysis, together with a proposal for a qualitative approach to economic problems arising in the building process. Qualitative reasoning formalisms needed for time-modelling will be presented in considerable detail.

Next, we will analyze the need to continually monitor the economic performance of buildings in use. This subject has been broached in various contexts throughout the book, most importantly in connection with production plans. Instead of concentrating on investment decisions pertaining to a single building or its component, we will focus on the total building stock controlled by an owner. Again, time is of the essence here. We will discuss some new approaches to monitoring the entire real property holdings underlying an economic process. In the last analysis, the economic process offers the only relevant context for economizing the use of building resources.

In keeping with the Austrian tradition that provides the foundation of the analysis throughout this book, we will in some sense disregard the orthodox division between microeconomics and macroeconomics. Using the individual building owner as the focal point of analysis does not preclude dealing with phenomena such as business and building cycles, traditionally classified in the field of macroeconomics. The economy as a whole of course influences each economic subject. Outside that context macroeconomic issues are in fact devoid of substance.

Although the reader will be alerted to the most important connections between this and preceding chapters, it should be noted at this juncture that some of these connections may require occasional detours into material already covered. The reader is therefore advised to reconsider those sections that pertain to the indeterminacy of economic processes and the attendant uncertainty of economic knowledge. A full comprehension of this subject would warrant an acquaintance with the literature cited, especially Hayek [1948], Mises [1966, 1976, and 1978], Shackle [1969 and 1972], O'Driscoll and Rizzo [1985], and Lachmann [1986]. This may sound like a tall order, but the reader should keep in mind that there can be no shortcuts to economic theory.

MEASUREMENT AND CHANGE

Economics has apparently become a bona fide quantitative science. In any well-regarded economic journal mathematical expressions in economic arguments prevail. The trouble is that most people faced by economic problems find these writings unintelligible. Members of the profession are aware of the growing gap between high theory and economic practice. An increasing number of economists believe that "over-mathematization" of economics threatens to make it utterly irrelevant to practical affairs of economic subjects. However, the issue goes beyond the degree of mathematization, to the very role of mathematics in economics. Mathematical models of economic phenomena are most often constructed so that their parameters can be estimated from empirical data using statistical procedures. Given the nature of economic processes, as well as the nature of economic thought about these processes, quantitative reasoning as such is actually in question.

It is often argued that measurement is a precondition for the development of a quantitative science. "Science is measurement," as the old aphorism goes. However, measurement implies invariant relations between things. In economics, measurement is inadvisable, not because of technical difficulties with measurement itself, but because no constant relations exist in economic affairs. What can be measured is already history, and history is by and large unrepeatable. In Mises' [1966: 56] words:

> The impracticability of measurement is not due to the lack of technical methods for the establishment of measure. It is due to the absence of constant relations. [. . .] Economics is not [. . .] backward because it is not "quantitative." It is not quantita-

tive and does not measure because there are no constants. Statistical figures referring to economic events are historical data. They tell us what happened in a non-repeatable historical case.

Therefore, according to Knight [1934: 236, quoted by Kirzner, 1976a: 100], "[i]f we accept the aphorism, 'science is measurement,' as a definition of science [...], then there is no such thing as 'economic' science [...]." In Chap. 2, we dealt with the impossibility of measuring capital. In Chap. 3, we argued that only backward-looking measures of capital are available to us. According to Hicks [1973: 34–35]:

> Practical measures of capital [...] are almost inevitably backward-looking measures; for the data from which they are derived belong to the past. It is only on what has happened in the past that we have *information*.

Such information can guide action only in the absence of change, or, more precisely, in the absence of expectations of change. From a subjectivist standpoint, "change" can only be defined with respect to the state of expectations [O'Driscoll and Rizzo, 1985: 81, following Hayek, 1948: 41]. In the present discussion, a similar problem arises. No matter how much information we may acquire about building costs and benefits in the past, it can never provide a firm basis for choice in a world devoid of constant relations between the parameters of building economy, that is, a world in which change is expected from the outset. Especially when dealing with remote time-horizons, we must rely on our imagination.

In some cases and under some circumstances, economics can help us predict the effects of definite measures or policies. It can answer the question whether definite means are adequate to attain the ends aimed at. However, such predictions can be only "qualitative"; again, they cannot be "quantitative" because there are no constant relations between economic variables [Mises, 1978: 67]. We will return to this subject shortly.

Because all real economic problems revolve around change—as perceived and anticipated by economic subjects—constant relations can be found only in areas that fall outside the domain of economics proper. An example of this would be purely technological issues that pertain to production processes. The very distinction between "quantitative" and "qualitative" economics is therefore meaningless. In a sense, economics is by its very nature qualitative. As we will soon see, this does not preclude rigorous reasoning. The basis for such reasoning can be found in deductive logic.

TIME, AUSTRIAN ECONOMICS, AND QUALITATIVE REASONING

Time, one of the essential economic categories in Austrian economics, is of utmost importance to building economics. A systematic approach to the study of building processes is inconceivable without this dimension of human action, as has been shown in the preceding chapters. However, it is important to emphasize that the Austrian conception of time goes well beyond the mere temporal extension of economic processes to the subjective experience of the passage of time.[2] We must understand time in the context of the other distinguishing aspects of Austrian economics.

What distinguishes Austrian economics from other schools of economic thought? Three features may be singled out: first, radical subjectivism regarding both human preferences and expectations; second, the proposition that time cannot elapse without knowledge changing, with the consequence that the future is unpredictable; and third, a distrust of all formalizations of economic behavior that neglect the very source of economizing—the mind of an economic actor. Of course, these features are intertwined.[3]

The interplay between time and change therefore permeates all aspects of Austrian economics. The autonomy of the mind precludes determinism because knowledge shapes action and action shapes the world. "The impossibility of prediction in economics follows from the fact that economic change is linked to change in knowledge, and future knowledge cannot be gained before its time" [Lachmann, 1977: 90]. Therefore, the indeterminate nature of the world precludes prediction, and this severely constrains all attempts at formalization.

What role can economic theory play in such a world? If quantitative prediction is outside the bounds of economic theory, then the primary task of theory is to classify—focusing on what *can* happen as a sequel to a particular situation [Shackle, 1972: 72, quoted by Lachmann, 1978b: 15–16]. Theories telling us what can happen are substantially different from theories pretending to tell us what *will* happen under a certain set of assumptions. Therefore, economics should become considerably more descriptive than it is at present, and economists should acquire the skills needed to classify, describe, and compare many possible situations [Lachmann, 1977: 264, and 1978b: 16].

The field of qualitative reasoning, a subfield of artificial intelligence, offers some tools for this kind of approach.[4] Qualitative reasoning formalisms, based on deductive logic, have been developed precisely to address the possible futures inherent in a given set of assumptions about the present state of a system. Even more important, qualitative reasoning formalisms address the cause-effect chains essential to the understanding of processes in general, and economic processes in particular. The cause-effect chains with which we are concerned in economics always involve decisions. More specifically, in each link of the chain the decision will be intermediate, connecting its cause and effect [Hicks, 1979: 88–89]. These processes can be modelled *via* sequences of transitions from one qualitative state to another, each of which represents a structurally distinct episode in a system's history. Clearly, such a framework may be useful in connection with multiperiod planning, introduced in Chap. 2.

We should bear in mind, however, that qualitative reasoning formalisms cannot provide more than a systematic framework or skeleton for analysis, the results of which must be interpreted and modified in light of our understanding of the residual interdependencies between economic categories. These interdependencies would be lost in an approach that requires strict logical independence of all categories. Nevertheless, qualitative reasoning formalisms in existence today already suggest an analytical framework that can accommodate a large number of possible futures.

THE NEED FOR QUALITATIVE REASONING

Qualitative reasoning about physical systems is at present almost synonymous with the so-called naive physics.[5] Of course, there is nothing naive about this field. It is naive only in relation to unrealistic expectations associated with the extant scientific orthodoxy. This area of study has arisen in response to real difficulties encountered by artificial intelligence researchers attempting to impart knowledge about the physical world to computers. Although of great value in our daily lives, commonsense knowledge about physical systems does not coincide with our scientific knowledge. Capturing knowledge about air-conditioning systems, elevators, or revolving doors, for example, requires a different kind of formalization. Modern science does not offer proper foundations for such an endeavor.[6]

This, *mutatis mutandis,* applies to the economic world—as exemplified by the intricacies of market processes—and to modern economic science. An expert system concerned with economic advice can hardly be constructed on the basis of economic principles found in standard economics textbooks alone. Commonsense knowledge about economic behavior, undoubtedly useful in our day-to-day affairs, requires a different kind of formalization. This is not to say, however, that commonsense knowledge can be captured without remainder.

We should pause for a moment to point out that economic theorems are qualitative in character. As we have seen in the preceding chapters, they concern relative directions of change in economic variables. Take for example the "laws" of demand and supply: if the price of a normal good increases, then the quantity demanded of that good per unit of time will decrease, while the quantity supplied per unit of time will increase, other things remaining equal. When these relationships are depicted in a two-dimensional graph with price on the vertical axis and quantities demanded and supplied on the horizontal axis, we obtain a downward sloping demand curve and an upward sloping supply curve. Few economists would dispute this statement about the two curves. A larger number of economists would disagree with the statement that the demand curve is generally concave with respect to the origin, for example, implying that the quantity demanded per unit of time decreases at a decreasing rate as the price increases. However, there is significant dispute regarding the specific shapes of demand and supply curves contained in mathematical expressions of various kinds. In short, the qualitative relationships are more robust than quantitative specifications of these relationships.

Returning to our argument, a large proportion of human knowledge is tacit, and thus it resists formalization of any kind.[7] Expert system development is also constrained by the nature of tacit knowledge, which cannot be captured *in toto;* however, it can be approximated to a degree, and qualitative reasoning may offer some guidance in this task. More important, qualitative reasoning allows us to formalize some subjectivist economic concepts.

In naive physics, the direction of change is a crucial idealization, and making inferences about the future is a crucial objective. The incompleteness of our knowledge forces us to think about alternative possible futures, rather than a single future.[8] This indeterminacy provides an essential link between qualitative reasoning and subjectivist economics.

There is a danger that "possible futures" will be understood mechanistically, as well-defined states containing the "single future" itself. Actually, we can only delimit a *class of outcomes,* because genuine indeterminacy involves an open-ended or unlistable set of possibilities [O'Driscoll and Rizzo, 1985: 4, 27, 71]. Furthermore, in economics we are facing genuine indeterminacy of human behavior, rather than mere incompleteness of our knowledge about the behavior of determinate physical systems.[9] Which of the possible futures will be realized depends on human action, and therefore on intersubjective interaction including both cooperation and conflict. As a consequence, qualitative reasoning formalisms are likely to be less appropriate in the domain of economics than in the domain of physics.[10]

NAIVE BUILDING ECONOMICS

By analogy with naive physics, a systematic application of qualitative reasoning formalisms to problems of building economy would lead to the development of naive building economics. Many ideas presented in this book in fact imply and require such an approach to economizing the use of building resources. The incompleteness of our knowledge and the genuine uncertainty associated with our conception of the future point at the need for tools and approaches that acknowledge the limitations of planning in the building realm.

What is naive building economics about? Let us begin by considering two basic requirements for its development. First, it should give the user of expert systems, decision-support systems, or other software tools dealing with building economy, a clear understanding of the underlying economic principles, not merely a justification of the rules-of-thumb used by experts. The formalization of knowledge in this field is in danger of becoming entangled in the idiosyncratic aspects of expertise.

Second, it should not run counter to commonsense reasoning about building economy that the user employs in everyday decisions. It should help the user in the process of assumption-making about the qualitative relationships between the key variables of a problem. The user's assumptions are based on his or her expectations about long-term economic trends and tendencies, both internal and external to the economic process in question. This is especially important for time-related phenomena involving structural change. Quantitative analysis may eventually be approached only on that basis.

There is no doubt that a program for developing qualitative reasoning in building economics is reductionist in nature. A significant proportion of information may thereby be lost. Quantitative analysis could not be replaced in its entirety, no matter how tentative our interpretation of its results might be. The proper place of qualitative reasoning is in the assumption-making stage of a project, during which most quantitative information is either missing or unreliable. The connection between assumptions and conclusions must be understood thoroughly before costly information gathering and processing can proceed to the quantitative stage.

The art of assumption-making must rest on firm logical foundations. That is precisely where the strength of artificial intelligence tools lies. Symbolic programming and logic programming are thus common synonyms for programming approaches developed in the field of artificial intelligence. Logic is needed to determine whether assumptions imply conclusions.[11] Indeed, an expert system development tool, and its inference engine in particular, may be thought of as means to ensure the logical consistency of the knowledge base built for a particular domain.

The role of qualitative reasoning in the assumption-making process is rather obvious; to ensure that the user understands the underpinnings and consequences of a particular choice. Commonsense notions about how the world works cannot be bypassed in this process.[12] Qualitative reasoning therefore focuses on notions that describe trends and tendencies, as well as expected changes in trends and tendencies, in the behavior of key variables. It also concerns causal relationships between the key variables, which require a special representation medium, that of deductive logic. Because causal relationships may lead to multiple futures, one of the main roles of deductive logic is to keep these futures distinct for further analysis.[13] The task of determining the causal chain of consequences of our actions is undoubtedly in the domain of assumption-making, much of which can be automated.[14]

Thus far little has been said to distinguish qualitative reasoning in building economics from qualitative reasoning in general. One of the central concerns of building economics must be that of modelling building processes. As long-lived capital assets, buildings must be modelled with special attention to dynamics. Change and time are inherently related.[15] In fact, most commonsense reasoning about building economy is dynamic in that it is continually modified through novel experiences.[16]

Building planning and design are cases in point. The very decision is shaped through a protracted process beginning with assumption-making. It is worth stressing that assumption-making is an essentially creative activity in which one endeavors to circumvent data availability and unreliability problems in early stages of planning and design. As Simon [1981: 171] argues, "[t]he heart of the data problem for design is not forecasting but constructing alternative scenarios for the future and analyzing their sensitivity to errors in the theory and data." Some expectations are invalidated in the process, while others become even more firmly entrenched as time-honored rules-of-thumb. The formalization of assumption-making about the dynamics of building is therefore fundamental to the endeavor proposed here, as it is a precondition for systematic knowledge acquisition and testing. Put differently, the very fact that buildings evolve through a rather lengthy process involving numerous decision sequences implies that learning is fundamental to the building activity, as well as that the underlying learning process needs to be anticipated and guided from the outset. However, it is worth emphasizing again that no analytical framework will ever lead to the removal of genuine uncertainty about the future.

TIME-MODELLING AND BUILDING PROCESSES

Clearly, time-modelling is an important aspect of qualitative reasoning in general. Time is modelled in terms processes, in which each separate time-interval is treated as structurally distinct, as an "episode." In other words, the very structure of the system under consideration may qualitatively change from one episode to another. This is crucial for thinking about buildings and building processes because distant time-horizons require that the expected structural change be explicitly included in the picture. In short, time-modelling is especially important when dealing with long-term phenomena, such as buildings. Research in this field should thus concentrate on building dynamics, in terms of qualitatively distinct states an economic model representing the building process may take under different conditions.

Let us introduce a simple example of time-modelling. It concerns real interest rate trend deduction, given the assumptions about the direction of change of the nominal interest rate and the inflation rate. By definition, real interest rate is equal to nominal interest rate minus inflation rate. So, if the nominal interest rate is increasing while the inflation rate is decreasing, we can readily deduce that the interest rate is increasing. However, if both the nominal interest rate and the inflation rate are increasing or decreasing, the deduction of the real interest rate trend will depend on the assumption about the relative trends of the two rates. Assuming that both the nominal interest rate and the inflation rate are increasing, but the inflation rate is increasing faster, then the real interest rate will be decreasing. A complete set of deduction procedures concerning this case can easily be realized using qualitative reasoning formalisms.

Now, let us consider an apparently more complex example of time-modelling based on the previous one. The trend deduction scheme already described may be used both with respect to the present and the future—say, a year from now—with the objective of determining whether we should build now or wait a year. The two episodes are structurally distinct. It can be shown that, everything else remaining equal, we should build now if the real interest rate is expected to be increasing in a year's time, and vice versa. Put differently, the outcome does not depend on the present behavior of the real interest rate, but only on its expected future behavior. Although such a deduction is almost trivial, the number of possibilities grows rapidly as the problem becomes more complex. A considerably more elaborate example of time-modelling, using qualitative reasoning formalisms to address the question whether to build sooner or later, is provided in the Appendix.[17]

When the user's assumptions are time-dependent it is especially difficult to address them quantitatively in the assumption-making process. The most obvious reason for this is that finding the appropriate quantitative values for the key variables may be prohibitively difficult, if not impossible. Moreover, trusting any numbers over long stretches of time may be foolhardy. In such cases qualitative reasoning offers an alternative approach.

TOWARD A RESEARCH PROGRAM
FOR NAIVE BUILDING ECONOMICS

This section does not comprise a definitive listing and justification of the tasks before us, but it is a step in that direction. In fact, some degree of vagueness is not only tolerable but desirable at present. Naive building economics depends on a collective effort.

The requisite interplay between economic principles and commonsense reasoning in the process of assumption-making about building economy cannot be achieved by adding together expert systems designed to work in narrow domains.[18] Several general propositions are important to emphasize in this context. First, this calls for a dense, thorough, and uniform formalization, that is, a richly intertwined network of readily available and implementable concepts. On the technical side, we need a decision-support system composed of a large number of interacting modules designed to tackle conceptually decomposable tasks. A successful decomposition is inconceivable without economic theory, however.

Second, this formalization should be especially sensitive to the intricacies with which commonsense reasoning treats time. Time is essential to building economics, conceived as a subfield of the theory of capital. Moreover, time-modelling in this field must accommodate the user's subjective approach to assumption-making. The future is open in the sense that there are multiple futures between which the user must choose as the building process evolves. However, the possibilities available for choice must first be identified and presented in a systematic fashion. This is a classification problem *par excellence*. It is here that economic theory will play its most important role, as we have already argued.

Third, the role of economic knowledge in assumption-making is not merely to tell the user what the best course of action is, but to consult, advise, and aid in the process of decision making, as well as to evaluate critically the user's proposals. We need to acknowledge not only the demand for a "user-friendly" interface, but also respect the user's knowledge, creativity, and responsibility. This is especially important in the context of building economics, because building planning, design, construction, and operation cannot be approached as singular events in space-time. We must recognize that the process potentially leads to novel, unpredictable choices.

Rapid developments in the fields of expert systems and decision-support systems will undoubtedly facilitate the development of the analytical framework for building economics. However, naive building economics requires a unified effort, rather than a multitude of independent software systems designed to address a specific problem. It is therefore likely that this field will initially find most fruitful ground in real estate and construction divisions, departments, and offices of private and public organizations that own building stocks of considerable size and complexity. Such organizations have the strongest incentive to manage their real property in a systematic fashion, because they are very much concerned with the coordination of all activities that affect the planning, development, design, construction, and management of their entire property holdings. We will turn to this topic shortly.

BEYOND LIFE-CYCLE COSTING

At present, the most developed area of building economics concerns investment criteria based on standard discounting principles. These principles have been borrowed from the fields of capital budgeting and benefit-cost analysis, concerning respectively private and public sectors, and applied to the specifics of the building field. The so-called life-cycle costing analysis is a case in point. Since the two energy crises in the 1970s, which spurred the development of building economics in general and life-cycle costing in particular, it has been standardized and systematized to a considerable extent. This is true especially in the public sector, where it is often mandatory in the project evaluation stage.[19]

However, life-cycle costing methodology should be further developed to monitor continually buildings in use. As we have seen in this book, and especially in Chap. 3, the emphasis on early stages of building planning and design is unwarranted. The primary objective of a comprehensive methodology of life-cycle analysis is to maintain an up-to-date picture of options available to the owner and/or the user throughout the life cycle of a building. A host of ostensibly minor day-to-day decisions involving building maintenance and replacement in the aggregate represent significant expenditures escaping adequate control. Of course, this argument holds for capital expenditures in general. According to a corporate treasury official of an American corporation, quoted by Bromiley [1986: 112]:

> The capital investment decision is the fundamental day-to-day decision this company makes. It's the most important decision this company makes. [. . .] It is the fundamental decision that is made on a day-to-day basis.

Note the repeated emphasis on the day-to-day character of investment decisions, as distinct from the prevalent conception of investment as a discontinuous and sporadic activity concerning new capital.

Even more important, considerable work yet remains to be done in the area of monitoring the entire building stock, that is, real property portfolio, managed by an owner. Many choices involve trade-offs between actions that span several properties: to reroof one building or to recarpet another, to increase the level of maintenance of one building or to replace the elevators in another, etc. Furthermore, many choices involve trade-offs between actions affecting all the properties owned: selection of a centralized computer system for preventive maintenance, determination of the overall reroofing budget for a particular year, etc.

One of the most important preconditions for these extensions of life-cycle costing is the development of owner-specific data bases including all relevant information about their individual buildings and the total building stock at their disposal. It is likely that both structure and content of such data bases will remain specific to each owner, as economic conditions relevant to real property differ significantly from one type of owner to another, one geographic area to another, as well as one interval of time to another. Nevertheless, the real property management approaches and the underlying information and knowledge will also share some features. The common features will be associated primarily with the type of activity in which the owner is engaged, type of ownership, and the size or complexity of property holdings involved.

REAL PROPERTY PORTFOLIO MANAGEMENT

There are several broad justifications for increased attention to real property management.[20] They are associated with the key problems facing the field. First, approximately one quarter of corporate assets in the United States are in the form of real property, buildings and land.[21] That percentage is likely to be somewhat higher in the public sector, because its capital structure tends to be less equipment-intensive. However, most organizations do not perceive themselves being in the real estate business. As a consequence, real property is by and large undermanaged. For instance, real estate and construction divisions, departments, and offices are typically perceived as cost-centers, rather than profit-centers. The value of real property thus tends to be neglected. When the relationship between cost and value of real property is recognized, its economic performance may be improved. An improvement in the management of real property is likely to have a significant impact on an organization's overall performance.

To avoid possible confusion, let us return briefly to the problem of monetary calculation—discussed in Chaps. 2 and 3—in the context of the value of real property. It should be emphasized here that monetary calculation is not the measurement of value; rather, it is an ordering and comparison of values. In Mises' [1976: 160] words: "[i]n the last analysis, economic calculation does not rest on the measurement of values, but on their arrangement in an order of rank." Of course, this is an essentially qualitative procedure. We should bear in mind that this argument pertains to costs also. As we have argued in Chap. 3, value and cost concepts are intrinsically related.

Second, at any given time-interval, the multitude of activities regarding the utilization and operation of real property typically requires significantly greater resources than do new investment projects. This is especially true of large, mature organizations. Although new investment decisions are carefully considered by the top-level management, decisions concerning the utilization and operation of capital goods are more often than not left to lower- and middle-level management. According to Drucker [1985: 68]:

> Almost every company has elaborate procedures for capital appropriations. [. . .] Most managements spend an enormous amount of time on capital appropriation decisions, but few pay much attention to what happens after the capital investment has been approved. In many companies there is no way of finding out. To be sure, if the new multimillion dollar plant gets behind schedule or costs a great deal more than was originally planned, everybody knows about it. But once a plant is on stream, its performance is rarely compared with the expectations that led to the investment.

The most likely reason for this situation is that new investment decisions are considerably more intelligible in nature. They involve well-defined projects, as well as clear-cut resource allocation choices. Also, the conventional capital theory tends to overemphasize the importance of decisions concerning new investment. The focus needs to shift to capital utilization and operation, as we have argued in the preceding chapters. An approach that renders the entire domain of real property utilization and operation comprehensible and manageable is therefore of great potential value to various types of organizations. We will return to this issue shortly.

Third, the communication between top-level executives and real property managers is generally poor. Organizational strategic objectives are rarely expressed in terms accessible to those who manage the real property needed for the realization of these objectives. A simple metaphor may be useful here: just as a fleet of ships requires overall strategy and coordination among its individual vessels, so too does a "fleet" of buildings. The communication gap might be closed by stimulating the information flow about real property throughout an organization. This may require a common language for everyone concerned with improving the economic performance of real property.

For all these reasons, there is an urgent need for research in real property management approaches and tools for private and public organizations that are not primarily concerned with real estate. This orientation is distinct from real estate development, insofar as it concerns the entire real property life cycle, as well as the entire real property portfolio. Both cross-sectional and longitudinal or time-series analyses of the portfolio are required for this task.

This research work should be geared toward the executives with responsibility for real property management in the context of broad staff functions, such as administration, operation, and personnel management. It should involve lower- and middle-level management issues only to the extent that they relate to problems faced by executives in terms of coordination, oversight, and guidance of these staff functions. In other words, we should address the management of facilities management, while facilities management itself should be of secondary importance.

It should nevertheless be emphasized that executives cannot really understand their function in this field without a detailed knowledge of activities reaching all the way down the management hierarchy to the level of operatives. One of the causes of real property undermanagement is that those who manage do not know in sufficient detail what is going on at the bottom of the hierarchy. Lack of knowledge about how buildings work, and how their performance can be improved, contributes to poor communication between various levels of management, as well as between the management and operatives.

BUILDING PERFORMANCE INDICATORS

Real property management focuses on buildings and land in use, but it also concerns the acquisition and disposition of properties, together with financing strategies associated with these activities—such as whether to own or lease particular pieces of real property. However, special emphasis is placed on buildings in use, and thus on building performance. Building performance can be gauged using a set of appropriately selected indicators, focusing on the match between the demand for and supply of building services. The values of these indicators vary over time, and the frequency of measurement of their values should therefore correspond to the rate of their variation.

Before we proceed, let us consider briefly several examples of building performance indicators. On the demand side, these indicators may include the outstanding requests for space measured in area per unit of time, and the number of employee complaints concerning different aspects of building performance per unit of time. On the supply side, they may include area per employee or per function, and various components of operating costs per unit of time and unit of area, or per function. Such ratios can be derived from the "raw" data that most organizations already collect on a regular basis.

In view of our argument concerning measurement in economics, we should emphasize that the role of building performance indicators is to *guide* managerial action, not to substitute for it by attempting to "automate" the process of decision making. Neither the definition nor the measurement of indicators can ever be perfect. Some skepticism regarding the validity of indicators as guides for action is therefore essential in their interpretation. Strictly speaking, the indicators can at best *indicate* where the managers' attention needs to be directed. This kind of guidance is nevertheless of great potential value because attention can be viewed as a major scarce resource in an information-rich world [Simon, 1982: 456]. In general, the greater the frequency of measurement, the more likely it is that indicators will indeed inform managerial action. A persistent shift in the value of an indicator whose magnitude has hitherto defied interpretation may signal significant change in economic conditions. Once the shift has been perceived, this indicator may be more readily linked to another indicator whose value has also changed in a systematic fashion. A better understanding of the underlying economic trends and tendencies may indicate the need to consider plan modification. In short, a systematic watch of building performance indicators offers many opportunities for discovery and learning. This is one of the main reasons why changes in indicator values over time are likely to provide more useful information to management than their magnitudes at any moment in time.

There are at least three reasons for continual measurement of building performance. First, measurement is needed to check whether the actual performance departs significantly from the desired performance. The range of acceptable variation of a particular indicator can be expressed in terms of its maximum and minimum acceptable values (standards or norms). If a particular indicator moves beyond these values, managers should be alerted and should make an attempt to diagnose and correct the situation. The next measurement of the actual value of the indicator will establish whether it has shifted in the desired direction, and whether further managerial action is required. Performance indicators are therefore meant to guide action, rather than simply measure the level of building performance.

Second, building performance may be measured to establish whether a change in policy has had the desired effect, irrespective of the standards or norms mentioned above. For instance, a change in the internal rent structure should be evaluated in terms of the desired effect on the organization's space use, such as the reduction in space hoarding for possible future use. In such cases policy precedes measurement. Of course, the question of precedence is not absolute in any sense, as all measurements and actions take place in the context of a continual measurement-action stream with many specific feedback loops.

Third, and most important, performance should be measured with the objective of gradually changing the character of the entire real property portfolio *via* continual managerial action bent on improving real property performance. Here, the objective is to shift the entire distribution of indicator values, that is, to improve upon the mean performance value, as well as to reduce the variation around the mean performance value. For example, an indicator can be based on a periodic questionnaire measuring the subjective satisfaction of building users with the overall performance of various buildings in the portfolio. This is illustrated in Fig. 5.1 in the form of frequency distributions of indicator values for the entire portfolio, measured at two different points in time. Clearly, this is a never-ending process involving continual incremental improvement.

We should note that "good" or "bad" performance is a relative notion. Referring to Fig. 5.1, the shaded area indicating the overlap between the two distributions shows that best performance in one period may be considered worst in another. Moreover, it is worth emphasizing that at any one time there will necessarily exist both "good" and "bad" buildings, represented by the two tails of a distribution.

The systematic learning process about the real property portfolio can inform several types of action available to the management. They relate to different phases of the building process. Those properties that perform best can be "replicated" by feeding information about their characteristics into the acquisition of new buildings, as well as design of new buildings and redesign of the existing buildings. Those properties that perform poorly may also be disposed of, if they do not promise to perform better upon redesign and rebuilding. The majority of properties, characterized by satisfactory performance, may be temporarily neglected, assuming that there are no indications that their performance is in danger of becoming unsatisfactory in the future. The so-called management by exception is based on the notion that the managers' attention should be directed toward a relatively small number of key decisions.

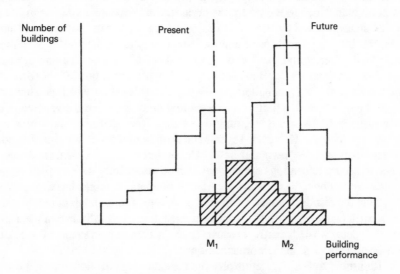

Figure 5.1 Strategic goals of portfolio management.

The key proposition underlying attempts to improve real property portfolio management is that one must establish a feedback loop between the performance of real property and managerial action, that is, between the demand for and supply of building services. As we have already seen, this involves a thorough understanding of cause-effect chains with managerial decisions as intermediate links. This applies to the decision making process connecting the entire real property management structure, including the top-level executives concerned. In a sense, incremental improvement that is always in line with changing organizational objectives is the "theory" behind real property portfolio management. It focuses on the provision of tools that would help change an organization's real property portfolio in the right direction, while refraining from determining what direction is right for an organization. The former aspect of this problem is by and large analytical; the latter aspect relates to the entrepreneurial function of top-level executives, which generally falls outside the domain of analysis.

SPACE ACCOUNTING AND PLANNING

Real property portfolio management must be based on an accounting system that captures changes over time. In other words, we need a dynamic accounting system, which can be used for space planning, evaluation, and monitoring of plan implementation. Here, we will briefly explore an important subsystem of the broader accounting system: the space accounting system. Useful space is undoubtedly one of the most important scarce resources at the disposal of an organization.

Let us observe an organization's use of space over an interval of time, in which all transactions are recognized at the end of the interval, say a month. Any particular square foot used for the organization's operations can be in one of the nine states denoted by circles in Fig. 5.2. It may enter the portfolio *via* three states, "build," "buy," and "lease and use" and it may leave the portfolio *via* three states, "sell," "lease and use," and "lease and hold." These five states attract most attention from top-level executives, as they affect the portfolio's size. We should note in passing that building demolition is not explicitly included here, although it may occur as a part of the rebuilding process. From the vantage point of the organization's space use, only two states, "own and use" and "lease and use" are productive. However, several states, although necessary, involve nonproductive space use, "own and hold," "lease and hold," etc. Vacant space is needed to secure relocations of personnel, and consequently the so-called vacancy rate must not be reduced below a certain minimum. In fact, a large proportion of space at the disposal of an organization is in one of these nonproductive states in any particular time interval. Top-level executives generally lack a clear picture of "space waste," partly because space accounting systems of the kind described do not exist in their organizations.

Returning to Fig. 5.2, let us observe the same organization during the next time interval. The same states pertain, but a particular square foot of space may

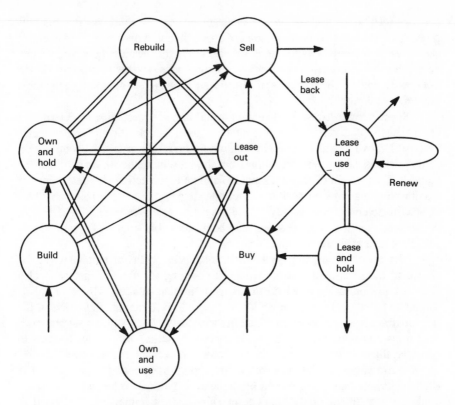

Figure 5.2   Space states and typical state transitions.

have made a transition from one state to another. The most typical state transitions are denoted by arrows. Clearly, the time interval covered by an accounting period ought to be short enough, so that a particular square foot does not pass through more than one state transition during any accounting period.

We should note that some state transitions depicted in Fig. 5.2 such as "build"–"rebuild" and "build"–"sell," may appear to be *prima facie* improbable, but they occur quite often in building practice as a result of changes in plans, discussed in Chap. 3. For instance, it is not uncommon for a warehouse to be transformed into an office building before the construction process has been completed, or that a new headquarters building is sold to another organization immediately upon its completion. Although such cases are often presented as regrettable "horror stories," they can be encountered fairly frequently in most large organizations, both private and public.

State transitions can also be represented along the time dimension. The number or the percentage of square feet in each state might be shown on the vertical axis, so that the changes in the amount of space in that state would be shown along the horizontal axis representing the time extension. The existing network techniques used in project management can be modified for this purpose. This is important in the context of space planning over time.[22] Again, we can never trust completely the figures presented in this fashion; however, we can be much more confident in their relative magnitudes or rates of change. This is often sufficient for the purposes of directing the managers' attention to the most pressing problems.

Space needs, that is, the demand for building services, must be determined on the basis of a plan. The present (and past) composition of space states described above must be known in addition to the future or desired composition of these states, so that the transition from the present to the future can itself be planned. The stream of resources available to an organization will determine the maximum rate at which transitions from one state to another can be made. That is, it will determine the supply of building services over time. Of course, a desired composition of space states may be inaccessible from the present state if sufficient resources are not available.

The space accounting system outlined here, with its emphasis on the choice of transition paths—from leasing to owning for use, for instance—also provides a monitoring mechanism for plan implementation. However, plans should be based on a couple of additional pieces of information, in particular, the information about the associated benefits and costs. Each space state and each state transition should reflect the associated net benefits. Among the benefits to consider, the most important are the use value and exchange value of real property. The use value may be measured by the replacement cost, for example. The exchange value may be measured by the assessed market value, which may include an assessment of the liquidation value. This framework, discussed in Chaps. 2 and 3, would offer a means to evaluate alternative courses of action available for space plan implementation. In general, we can say that the state transition path that maximizes the present value of net benefits should be selected for implementation.

Such a framework might be used to evaluate the consequences of alternative scenarios for an organization's real property portfolio. The space accounting system should therefore be embedded in a decision-support system facilitating scenario development exercises. Allowances for risk and uncertainty can be added to this picture. For this purpose, Monte Carlo simulation techniques might be used, assuming sufficient knowledge about probability-density functions associated with the time-horizons, costs, and benefits of particular courses of action.

MANAGEMENT OF ADAPTABILITY AND FLEXIBILITY

At this stage of the argument it is instructive to return to the problem of adaptability and flexibility, discussed in several places in the preceding chapters. We will consider this topic in the context of real property portfolio management tools and approaches. Because one of the key concerns of real property portfolio management is the maintenance of a continual capability to adapt to changing economic conditions, the structure of building performance indicators should correspond to the structure of the portfolio itself. The portfolio should be seen not as loose collection of buildings and parcels of land, but as a structure informed by a shared mission.

Therefore, the management of adaptability and flexibility should not be conceived narrowly, that is, in connection with individual buildings only. This problem arises on at least two additional levels, that of a cluster or group of buildings, and that of the portfolio as a whole. Furthermore, on all three levels there are three distinct aspects of the problem: physical, organizational, and financial.

The entire portfolio can be thought of as a structured collection of buildings, together with the associated parcels of land, as well as land available for future development. Some are "stand-alone" buildings distributed over a wide geographic area. Other buildings are grouped into clusters dedicated to a particular shared purpose or mission. These clusters may be concentrated in the same geographic area but on dispersed parcels of land, or they may form well-defined campuses on a single parcel or several contiguous parcels of land. Such a collection of buildings is diagrammatically illustrated in Fig. 5.3. In this example, there are two manufacturing campuses that include several manufacturing buildings (M), manufacturing-support offices (O), research laboratories (L), and warehouses (W); the headquarters complex concentrated in one geographic area, including several headquarters offices (H), a sales office (S), and a research laboratory; and a number of sales offices and a research laboratory distributed over a wide geographic area. Each building is represented by a small circle; manufacturing campuses are denoted by larger circles; the headquarters complex is represented by a dashed circle, indicating that buildings are located on dispersed parcels of land; and the largest circle denotes the portfolio as a whole. The real property management organizational structure typically corresponds to such a portfolio structure. In other words, each portfolio component is entrusted to the care of a particular manager supported by a staff.

Problems of adaptability and flexibility are quite distinct on different management levels, although they are of course interrelated. On the level of individual buildings physical issues predominate. They include, for example, column spacing and bay sizes, ceiling heights, number of floors and area per floor, horizontal and vertical expandability in terms of the site constraints, durability of main building components, etc. As the available options are constrained by the site, adaptability and flexibility tend to be perceived in terms of the physical alternatives available.

On the cluster level, physical issues include the site layout, overall expansion strategies, durability of entire buildings, physical connections between adjacent buildings, pedestrian and vehicular access, etc. Although physical issues are also very important at this level, the emphasis tends to be on organizational

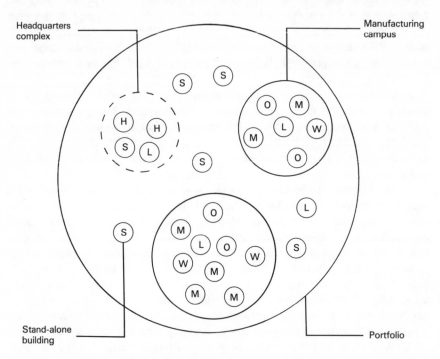

Figure 5.3   Real property portfolio structure.

issues. More specifically, management of adaptability and flexibility may involve the overall spatial distribution of various offices, departments, and
divisions that require contiguous space for their operations. Organizational units
that outgrow a particular building can be moved to a larger building on the same
campus, or a building that has been expanded for this purpose. Some space may
be provided in temporary buildings, which are adaptable and flexible in the sense
that they can be more readily relocated or removed if a major campus restructuring is required. On this level, adaptability and flexibility are likely to be perceived as less constrained by individual buildings and their physical
characteristics than by the campus itself.

Financial issues tend to predominate on the portfolio level, although they
undoubtedly play an important role on all levels. For example, the portfolio
level involves broad policies regarding the mix of owned and leased facilities
that corresponds to contingency plans for capital restructuring in case of major
changes in economic conditions. In particular, large organizations tend to own
the manufacturing campuses, research facilities, and most of their headquarters
offices, whereas they tend to lease a large proportion of their sales office space.
A building located in a manufacturing complex cannot be disposed of readily,

but a sales office building can. An organization can respond to geographic shifts in the demand for its goods and/or services by acquisition of new properties and disposition of old ones. Of course, some "physical" issues such as the geographical distribution of various facilities, are also very important on the portfolio level.

We should note briefly that not all the space state transitions depicted in Fig. 5.2 are equally applicable to all portfolio components shown in Fig. 5.3. For instance, it is highly unlikely that a manufacturing building will be sold or leased out, whereas this is not uncommon for sales office buildings. The former is a part of a campus; the latter is constrained neither by interaction with other buildings, nor by the contiguity of land parcels. In short, the possible courses of action differ for buildings in different parts of the portfolio.

Real property portfolio management tools and approaches discussed in this chapter ought to provide a coherent picture of adaptability and flexibility on all three levels, as well as in terms of all three aspects outlined here. The alternative courses of action available to top-level management should correspond to the options available at all other management levels. The entire real property management organizational structure should be conversant with the behavior of building performance indicators introduced in preceding sections. Again, the indicator structure should capture the essential characteristics of the portfolio structure, so that individual buildings can be compared in terms of their type, location, and interaction with other buildings in their proximity.

TOWARD A RESEARCH PROGRAM
FOR REAL PROPERTY PORTFOLIO MANAGEMENT

The value of performance indicators will tend to fluctuate over time. By and large, these fluctuations and the underlying trends will be of greater significance for real property management than the question of "satisfactory," let alone "best," indicator values. Although in some cases we can indeed state the acceptable range of variation of a particular indicator, in most cases such a determination can be made only on the basis of accumulated experience. In general, direction and relative magnitude of change of a particular indicator are likely to be of greater interest to the management than the specific magnitude of the indicator value.

Some performance indicators fluctuate with greater frequency than others in a given time interval. The frequency of measurement should correspond to the rate of fluctuation. Some measurements should be made once a year, while others should be performed quarterly, monthly, or even weekly.

However, the number of measurements, and thus the cost of measurement per unit of time, should be kept to a minimum. There are two aspects to this task. First, the number of performance indicators should be as small as possible. Organizations already have the bulk of information needed for the tools proposed here. Only a small proportion of it should be collected and maintained anew. Here a conceptual framework is indispensable for selecting most useful performance indicators. It should focus on capital utilization and operation, rather than the appropriate level of new investment. The substantive foundation for the development of the conceptual framework suggested here comes from the theory of capital as applied to real property in general and buildings in particular.

Second, measurements should be made using statistical procedures, which permit valid probabilistic inferences about the entire real property portfolio on the basis of relatively small samples. Again, efficient sampling cannot be achieved without a coherent methodological framework. Such a framework comes from the techniques that have been developed in the fields of industrial engineering and statistical quality control.[23]

Now it should be emphasized that these general propositions cannot be directly applied to every specific case. Implementation will depend on the nature of an organization and on its organizational structure. In particular, each organization will have a specific organizational goal and structure in the area of real property management itself. These characteristics will influence the way in which decisions are made and carried out, as well as the way in which information flows through various layers of management.

Some performance indicators will be useful to many types of organizations, because some problems are quite general in nature. One such problem is that of space-efficiency (in the facilities management field this is known as the space utilization problem). It might be measured in terms of square feet per unit of output (or per function) per unit of time. Most organizations would be interested in reducing "space waste" if they could get a handle on an effective approach to the problem. Management of space waste may in some cases require more than better facilities management; it may lead to reconstruction of a facility, or even its demolition and replacement by a new facility. However, it is generally worthwhile to try to improve the space utilization before recommending new construction. A well-constructed space-efficiency indicator should guide the decision-making process toward continual improvement, until the limits encountered become insurmountable and a capital decision must be made.[24]

Different organizations will generally need different indicator systems, however. One reason for this is that some organizations undergo much more rapid change than others. Another reason is that organizational missions differ considerably, and that this is reflected in the nature of their space needs. As a consequence, the frequency of measurement of a particular indicator may also differ across organizations.

In this context, our general approach should be flexible enough to permit different implementation strategies adequate to each case. At the same time, the implementation strategy adopted should preserve the coherence of the general framework underlying each instance of implementation. It is here that we would benefit most from the development of decision-support systems that simultaneously provide a comprehensive framework for real property management and offer a wide variety of implementation strategies. Such software systems should have the capability to accept information from several data bases available to an organization, as well as to adjust to the inevitable changes in these data bases, as they evolve in time.

The technology required for the decision-support systems of use in this field must therefore be versatile and flexible, and should permit and stimulate a learning-by-doing approach to the development of these tools. Excessively large and ambitious software systems that require long research and development periods before they can be tested and used, are not appropriate for real property portfolio management. Such systems tend to come on line too late. They also tend to be difficult to modify. "Quick is beautiful," as Dyson [1988: Chap. 8] argues concerning technology in general.

The qualitative reasoning formalisms already discussed in this chapter are likely to play an important role in the construction of decision-support systems for real property portfolio management. Such formalisms will be especially useful for combining the information about directions of change of interacting performance indicators made available by the system, as well as for constructing the causal chains that connect managerial decisions and their effects in terms of performance indicators. Of course, the art of assumption-making will become crucial in interactive exercises between an executive and his or her decision-support system, as the validity of inferences generated by the system will depend on the correctness of the assumptions made by the executive. These assumptions are inherently speculative, some will be validated and others invalidated by subsequent events.[25]

As was already argued in the preceding sections, the primary objective of research in this field should be to provide approaches and tools that facilitate the formation and maintenance of a feedback loop between real property performance across the portfolio and managerial action. This will open the road toward continual incremental improvement of real property performance, guided by the ever-changing objectives and structure of an organization.

CONCLUSION

As an emerging discipline, building economics requires the concerted effort of many people. This book can be thought of as an invitation to this collective endeavor. Although the pressing day-to-day work of building economists cannot wait for the full development of its theoretical foundation, without such a foundation a good portion of their work is likely to remain disjointed and even misguided. To paraphrase Drucker's [1985: 109, 111] admonition about the management of one's own resources, pressures always favor the past, while we should pick the future. The theory of building economics is akin to capital investment—it will make us more productive, but at the cost of a deferment.

It should be emphasized again that this chapter is not meant to be definitive or authoritative. It rests on my own perception of the field and its prospects for development. Nevertheless, I believe that its underlying imperative is cogent: The development of software tools to aid the decision-making process concerning the best use of buildings and land cannot proceed without economic theory. Propositions from building economics, as a branch of the theory of capital, should be embedded in the structure of these tools. The specific knowledge associated with the particular circumstances of time and place facing the user in practice will find a proper place only in this structure.

The proposals offered here are therefore both very demanding and quite modest. Demanding, because parallel development of theoretical underpinnings and practical applications is a formidable task in a field dominated by excessive pragmatism of the building professions. Modest, because we can proceed in small steps, formalizing and enhancing our knowledge *via* a feedback loop, as we proceed from project to project. This is as it should be, however. On the strategic level we should be bold, while proceeding cautiously on the tactical level.

Finally, I would like to stress that building economists will have to continue dealing with an enormous amount of detail characteristic of the building process. The danger here is that it is easy to get lost in the enormous amount of information required to manage real property well. Still, an economic understanding of the intricacies of the building process makes building economics indispensable. No other branch of economics concerns itself with the nuts and bolts of the building realm. To maintain a clear picture of this rich and sometimes confusing domain, and to gradually sift the important from the unimportant, theory is needed as a road map and a guide. This book has been written to fill that need. Only the reader can judge how well it has succeeded.

APPENDIX: AN EXAMPLE OF QUALITATIVE REASONING APPLIED TO TIME-MODELLING

We often face the choice of providing for a future need before or after it arises. The choice is between a known expenditure now, and a contingent expenditure later. The main element of uncertainty lies in the period of time likely to elapse before the need must be satisfied. In formulating the problem we must make this period the focus of analysis.

Let us consider a problem of this kind, adapted from Stone [1980: 82–85]. Suppose that a firm needs an office building of size A for its present needs, but that a larger size, B, will be needed sometime within ten years. The construction process is about to begin. According to the contractor's cost estimate, providing for size B initially will cost six million dollars more than for size A. The cost of increasing the size to B after completion of the building is estimated to be ten million dollars. For a building of size A the annual upkeep cost is estimated to be

$400,000, while for a building of size B these costs would be $750,000. The rate of discount (R) is taken to be 7 percent net. The respective cost escalators for upkeep (U) and construction (C) are both assumed to be 2 percent net. The value of the additional space to the firm is not included in the picture. Is it more economical to build size B initially or subsequently?

The answer can be found by standard discounting procedures of life-cycle costing analysis (see Stone [1980], for example). In this case it is instructive to concentrate on the period of time (T) required to reach the break-even point (E), at which the costs of initial and subsequent provision of size B are equal. Given our assumptions, if we can be reasonably sure that the need for size B will arise in less than five to six years, we should provide size B initially, and vice versa. At the break-even point we should be indifferent between the two options.

At this stage uncertainty can be introduced most simply into the analysis by varying the problem parameters. For example, we may assume that R will drop to 5 percent net; that U will increase to 5 percent net; and that C will remain stable at 2 percent net. Under these assumptions, the break-even point shifts to the right and T increases: we should provide for size B initially if we are reasonably sure that the new need will arise within seven to eight years. Both sets of calculations are presented graphically in Fig. 5.4 (full line denotes initial and dashed line subsequent provision of size B).

This simple example illustrates a broader class of such questions, and emphasizes the aspects of the problem requiring greatest attention. In this case, the time elapsed before the future need arises is the key element in the analysis. Period T and the sign of its first derivative (denoted in qualitative calculus by DT) can be addressed by means of qualitative reasoning formalisms.

Before we proceed with the example, we should introduce some fundamentals of qualitative calculus. Following Forbus [1985], we will distinguish between the sign and magnitude of a quantity. Let the magnitude of a quantity be denoted by A. Its sign will be denoted by [A]. Let the magnitude of the derivative of A with respect to time be denoted by dA/dt. Its sign will be denoted by $[dA/dt] = DA$. Signs can take on the values $-$, 0, and $+$. $[A] = +$ means that $A > 0$, that is, that A is positive. $DA = +$ means that $dA/dt > 0$, that is, that A is increasing. We sometimes need to combine sign values across addition. Table 5.1 illustrates the algebra used for $[A + B]$ and $[A - B]$. Note that a similar formalism is used to combine sign values across multiplication.

The cases marked by N1 require additional information to determine the sign value of $[A + B]$ and $[A - B]$. This is called disambiguation. For example, if $[A] = -$ and $[B] = +$, and we learn that $A > B$, then $[A + B] = -$. The sign of the sum in that case takes on the sign value of $[A]$. Additional information may be supplied only when it is needed to resolve ambiguities.

A similar procedure can be used to determine the ordering relationship between two quantities given the sign of their derivatives over some interval of time. Table 5.2 illustrates this operation for $A > B$ and $A = B$. Note that there is a continuity requirement that all order relations must go through equality.

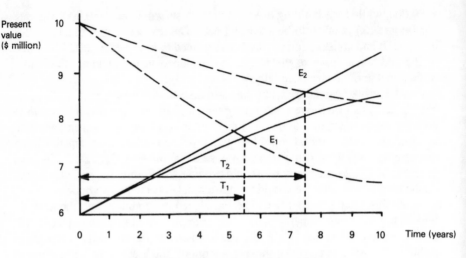

Figure 5.4 Discounted costs of alternatives.

TABLE 5.1   Sign Values Across Addition

	[A + B]				[A − B]		
	[B]				[B]		
[A]	−	0	+	[A]	−	0	+
−	−	−	N1	−	N1	−	−
0	−	0	+	0	+	0	−
+	N1	+	+	+	+	+	N1

N1: if A > B then [A]
    if A = B then 0
    if A < B then [B]

Therefore, the movement from A > B to A < B must go through A = B, that is, must have two episodes.

Now we can return to our example. There are nine (3 × 3) possible configurations of the curves shown in Fig. 5.4, depending on the relative magnitudes of R, U, and C. These configurations, or qualitative states, are shown in Fig. 5.5. Note that state S1 can be thought of as the "normal" state, which is likely to prevail most of the time. The other states represent special conditions and short-term profit opportunities or threats of loss. For instance, the building itself appreciates in value only when R < C, other things being equal. Note also that in states S1 and S9 it is not necessary that U = C (U > C and U < C are also possible).

TABLE 5.2   Derivatives and Inequalities

	A > B				A < B		
	B				B		
A	–	0	+	A	–	0	+
–	N2	=	=	–	N4	<	<
0	>	>	=	0	>	=	<
+	>	>	N3	+	>	>	N5

N2: if $dA/dt > dB/dt$ then = and > otherwise
N3: if $dA/dt < dB/dt$ then = and > otherwise
N4: if $dA/dt > dB/dt$ then <
    if $dA/dt = dB/dt$ then =
    if $dA/dt < dB/dt$ then >
N5: if $dA/dt > dB/dt$ then >
    if $dA/dt = dB/dt$ then =
    if $dA/dt < dB/dt$ then <

The system in question passes from one qualitative state to another, as the relative magnitudes and signs of dR/dt, dU/dt, and dC/dt change. For instance, the case shown in Fig. 5.4 represents the transition from state S1 to state S4. Each transition or change of state can be associated with a structurally different episode in the system's history. Note that states S2, S4, S5, S6, and S8 are "instantaneous" states, included here to satisfy the continuity requirement. The system can be in these states only instantaneously, while the remaining states may involve longer transition intervals.

Let us take a very simple example of a sequence of state transitions. Suppose that the system is presently in state S1, that DR = –, and DU = DC = 0. Possible sequences of qualitative state transitions in this example are shown in the form of a lattice in Fig. 5.6. Instantaneous states are denoted by squares, while the states involving longer transition intervals are denoted by circles. Note that transitions between instantaneous states S2 and S6, and S4 and S8, are logically possible but highly unlikely; the same holds for the transition from the initial state, S1, to the terminal state, S9, via the instantaneous state S5. Assuming that the stated conditions would prevail indefinitely, the system would reach state S9 either *via* state S3 or S7. This would depend on the relative magnitudes of U and C.

Now, the main difficulty we are facing here is that dT/dt is simultaneously determined by dR/dt, dU/dt, and dC/dt. Space limitations prevent investigating all the conditions under which transitions can take place. We will concentrate instead on the effect of state changes on the key parameter of choice—time

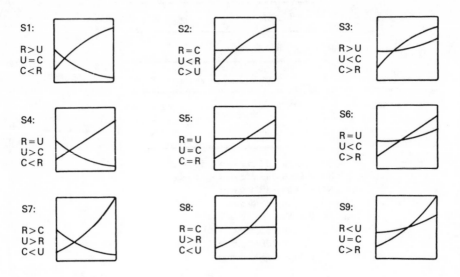

Figure 5.5   Possible qualitative states.

elapsed before the future need arises. That is, the analysis will focus on the direction of change of T.

Consider an expert system module that would inquire about the relative magnitudes of R, U, and C that the user believes are true at present. The system would inquire whether R, U, and C are increasing, decreasing, or stable. On the basis of this information the system would then deduce the direction of change of T. An increase of T would suggest that size B should be built initially, and vice versa. If the system were unable to make such deductions on the basis of the information available to it, it would inquire about the *relative* rates of change of R, U, and C, until all the remaining ambiguities were resolved.

How could this be accomplished when we are operating with incomplete or weak information? It is easy to show that dT/dt is inversely proportional to dR/dt and dU/dt, and directly proportional to dC/dt. Note that T decreases monotonically as R and/or U increase, but increases monotonically as C increases. It can also be shown that dT/dt is most sensitive to dR/dt, less sensitive to dC/dt, and least sensitive to dU/dt. The counterpart of functional relationships in standard calculus are qualitative proportionality (q-proportionality) relationships in qualitative calculus. Two quantities are directly q-proportional if their qualitative derivatives have the same sign value; they are inversely q-proportional if their qualitative derivatives have different sign values, excluding zero. The following relationships hold in our case:

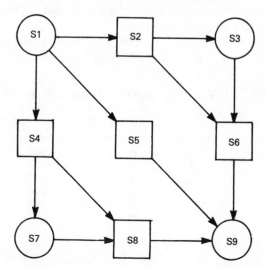

Figure 5.6 Possible sequences of state transitions.

(A) for DU = DC = 0: if $\begin{cases} DR = - \\ DR = 0 \text{ then} \\ DR = + \end{cases}$ $\begin{cases} DT = + \\ DT = 0; \\ DT = - \end{cases}$

(B) for DR = DC = 0: if $\begin{cases} DU = - \\ DU = 0 \text{ then} \\ DU = + \end{cases}$ $\begin{cases} DT = + \\ DT = 0; \text{ and} \\ DT = - \end{cases}$

(C) for DR = DU = 0: if $\begin{cases} DC = - \\ DC = 0 \text{ then} \\ DC = + \end{cases}$ $\begin{cases} DT = - \\ DT = 0. \\ DT = + \end{cases}$

Returning for a moment to the example shown in Fig. 5.4, we see that in that case we can easily deduce that DT = +. However, that holds for the terminal state S9, while we cannot tell much about DT in the transition process itself. This requires additional information.

Now, the above approach offers a qualitative counterpart to partial derivatives in standard calculus. Table 5.3 thus shows that we can deduce the sign value of DT directly in 7 out of 27 (3 x 9) possible combinations of sign values of DR, DU, and DC. A direct deduction is possible in four additional cases, since DR and DC, as well as DU and DC, are inversely q-proportional. These are the cases where DR and DC, or DU and DC, have the opposite sign values, while DU = 0 or DR = 0, respectively. Therefore, in 16 cases there remain ambiguities about DT, denoted by a question mark, that require additional information about the relative magnitudes of R, U, and C, as well as the signs and relative magnitudes of *dR/dt*, *dU/dt*, and *dC/dt*.

Let us suppose, for example, that DR = −, DU = 0, and DC = +, which corresponds to case 20 in Table 5.3. In this case we can readily deduce that DT

TABLE 5.3   Combining Sign Values of DR, DU, DC, and DT

	DR	DU	DC	DT		DR	DU	DC	DT		DR	DU	DC	DT
1	+	+	+	?	10	+	0	0	−	19	−	+	0	?
2	0	+	+	?	11	−	−	+	?	20	−	0	+	+
3	+	0	+	?	12	−	+	−	?	21	0	0	−	−
4	+	+	0	?	13	+	−	−	?	22	0	−	0	+
5	−	+	+	?	14	0	0	0	0	23	−	0	0	+
6	+	−	+	?	15	+	0	−	−	24	0	−	−	?
7	+	+	−	?	16	0	+	−	−	25	−	0	−	?
8	0	0	+	+	17	0	−	+	+	26	−	−	0	?
9	0	+	0	−	18	+	−	0	?	27	−	−	−	?

= +. Therefore, there is no need for so-called disambiguation. But if we suppose that DR = −, DU = +, and DC = 0, which corresponds to case 19, additional assumptions will be needed to resolve the ambiguity.

Let us suppose now that R > U, U = C, and C < R, that is, that the system is starting from state S1; that *dR/dt* < 0, *dU/dt* > 0, and *dC/dt* = 0; and that *dR/dt* < *dU/dt*. These assumptions correspond to case 19 in Table 5.3, and to the case shown in Fig. 5.4. Using the qualitative calculus operations presented in Tables 5.1 and 5.2, and assuming that U > C, it can be deduced that the system would first make a transition from state S1 to S4 to S7, and then from state S7 to S8 to S9. It can be shown that in the first few episodes DT = −, while in subsequent episodes DT = +. Period T would thus eventually increase if the current assumptions continue to hold, though this would be unrealistic to expect over an extended period. The sequence of qualitative state transitions in this example is shown in Fig. 5.6. Similar deductions can be made about the other cases.

The resolution of ambiguities involves a gradual reduction of the number of possibilities as new information becomes available. New information may involve postponed decisions. "In a very simple form: if you decide to make up your mind only next Saturday, *not before*, on how to spend the weekend, you cannot possibly know *now* what you will do next Sunday" [Georgescu-Roegen, 1971: 335]. In such cases ambiguities are inherent in the decision-making process itself. Ambiguities are not flaws in the process, but indicators of incomplete or weak information. Quantitative approaches, in contrast, suffer from artificial exactness and accuracy. They also imply that all the information concerning the future is already available in the present. This pertains especially to deterministic analytical approaches, which dominate the building professions and business practices. They convey a false sense of security through misplaced numerical precision.

The rough edges remaining in this presentation are not necessarily due to limitations of qualitative reasoning. Because of space limitations we have focused on the bare essentials. We may conclude that deductive logic is in some cases sufficient in cost-modelling exercises. When cost data is unavailable or untrustworthy, qualitative reasoning formalisms may prove indispensable. This is especially true in the early stages of the building process, when the user must understand the logical implications of his or her assumptions.

NOTES

1. For an overview of expert systems and decision-support systems, respectively, see Waterman [1986] and Mittra [1986], for example.
2. As Lachmann [1978b: 1] argues, "[t]time, as the dimension of the interval between input and output, is important, but is not all-important." In their elucidation of the Austrian conception of time, O'Driscoll and Rizzo [1985: 52–70] distinguish between Newtonian and Bergsonian or real time, where the former is symbolized by movement along a line, and the latter represents the subjective experience of the passage of time. They follow Bergson [1910] in further distinguishing between the static subjectivist concept used in planning and a dynamic subjectivist concept of continuous flow of novel experiences. As O'Driscoll and Rizzo [1985: 60] argue, "[w]e cannot experience the passage of time except as a flow; something new must happen, or real time will cease to be." It is interesting to note in passing that the title of their book, *The Economics of Time and Ignorance,* probably owes its origin to Keynes' [1964: 155] pronouncement that "[t]he social object of skilled investment should be to defeat the dark forces of time and ignorance which envelop our future." Of course, this is the perspective of this book, as well.
3. In Lachmann's [1978b: 1–2] words:

> The first, and most important, feature of Austrian economics is a radical subjectivism, today no longer confined to human preferences but extended to expectations. [. . .]
>
> Secondly, Austrian economics displays an acute awareness of the many facets of time that are involved in the complex network of interindividual relations. [. . .] To Menger, time was, in the first place, the dimension in which the complex network of interindividual relations presents itself to us. Austrian economics has retained and cultivated this Mengerian perspective. Time is the dimension of all change. It is impossible for time to elapse without the constellation of knowledge changing. But knowledge shapes action, and action shapes the observable human world. Hence it is impossible for us to predict any future state of this world.
>
> The third feature of Austrian economics, a corollary of subjectivism and awareness of the protean character of time, is a distrust of all those formalizations of

economic experience that do not have an identifiable source in the mind of an
economic actor.

It should be noted that the second proposition—that there is no period of time in
which change does not take place—is of considerable vintage. Newton-Smith [1980:
13–14] refers to it as Aristotle's principle, which he contrasts with the Platonist commitment to the possibility of time without change.

4. For an introduction to qualitative reasoning about physical systems, see Bobrow
   [1985]. See Bon [1986b and 1987] for applications of qualitative reasoning formalisms to problems of building economics.

5. This notion was introduced by Hayes [1979].

6. According to De Kleer and Brown [1985: 13]:

   Naive physics concerns knowledge about the physical world. It would not be such
   an important area of investigation if it were not for two crucial facts: (a) people are
   very good at functioning in the physical world, and (b) no theory of this human
   capability exists that could serve as a basis for imparting it to computers. Modern
   science, which one might think should be of help here, does not provide much help.

7. O'Driscoll and Rizzo [1985: 104] recognize that "[i]n all areas of human endeavor,
   individuals employ knowledge that either they are not aware they possess or they cannot characterize precisely enough to communicate to others." An example of this are
   the particular circumstances of time and place. (See note 6 in Chap. 3.)

8. As Forbus [1985: 89] writes, "we usually must work with incomplete information, so
   we can only generate descriptions of possible futures, rather than a single future."

9. It should be noted that this distinction by no means applies to "mind and matter" in
   general, but only to "macroscopic" physical systems. As Dyson [1988: 8] points out,
   the sharp distinction between mind and matter is no longer tenable in physics:

   When we examine matter in the finest detail in the experiments of particle physics,
   we see it behaving as an active agent rather than as an inert substance. Its actions
   are in the strict sense unpredictable. It makes what appear to be arbitrary choices
   between alternative possibilities. Between matter as we observe it in the laboratory
   and mind as we observe it in our consciousness, there seems to be only a difference
   in degree but not in kind.

   According to Dyson [1988: 297], "[i]t appears that mind, as manifested in the
   capacity to make choices, is to some extent inherent in every electron."

10. According to Forbus [1985: 164], "the features which make qualitative models useful
    for physical reasoning [. . .] should be useful in other domains, especially in domains
    where numerical data is unreliable or hard to come by." However, Forbus [1985:
    164] cautions that

    There seems to be no real agreement on what mathematical descriptions are ap-
    propriate in economics, hence it will be hard to judge whether a qualitative model
    is correct. In addition, the very structure of the domain can change with time; for

instance, the tax code can change. These factors make modeling economics much harder than modeling physics.

Some economists have approached the problem of prediction in a way that is in many respects similar to qualitative reasoning. For a discussion of prediction in economics, and especially the prediction of the *direction* of real-world changes, see O'Driscoll and Rizzo [1985: 79–88]. Their analysis focuses on prediction in the context of plan coordination, that is, plan compatibility concerning activities of different economic agents. It should be noted that Austrian economists generally follow Hayek [1948: 41] in thinking of equilibrium in terms of plan compatibility [O'Driscoll and Rizzo, 1985: 80]. Of course, equilibrium is a fundamental concept of economic analysis.

11. As Kowalski [1979: i] writes, "[logic] investigates whether assumptions imply conclusions, independently of their truth or falsity and independently of their subject matter."

12. Thus De Kleer and Brown [1985: 8] write:

Artificial intelligence and (especially) its subfield of expert systems are producing very sophisticated computer programs capable of solving tasks that require extensive human expertise. A commonly recognized failing of such systems is their extremely narrow range of expertise and their inability to recognize when a problem posed to them is outside this range of expertise. In other words, they have no commonsense. [. . .] The missing commonsense can be supplied, in part, by qualitative reasoning.

13. As Hayes [1979: 246] put it, "'[c]ausality' is a word for what happens when other things happen, and what happens, depends on circumstances."

14. The proposal to automate the assumption-making process should not be construed as inimical to reasoning as such. In a beautiful passage, Whitehead [1969: 41–42] writes:

It is a profoundly erroneous truism, repeated by all copy-books and by eminent people when they are making speeches, that we should cultivate the habit of thinking of what we are doing. The precise opposite is the case. Civilization advances by extending the number of important operations which we can perform without thinking about them. Operations of thought are like cavalry charges in a battle— they are strictly limited in number, they require fresh horses, and must only be made at decisive moments.

It should be noted that he made this argument in the context of a discussion about mathematical symbolism. According to Whitehead [1969: 41], "by the aid of symbolism, we can make transitions in reasoning almost automatically by the eye, which otherwise would call into play the higher faculties of brain." Indeed, this is the precise aim of qualitative reasoning.

15. According to O'Driscoll and Rizzo [1985: 57], "[c]hange, rather than mere uncertainty, is the *true* effect of time."

16. As O'Driscoll and Rizzo [1985: 63] write:

Novel experiences acquired in planning or acting are significant only to the extent that they engender plan revisions or alterations in the course of action. Thus these experiences must be connected with changes in the stock of knowledge. Knowledge, unlike pure experience, has applicability beyond the individual case that gave rise to it and thus can affect the future.

17. The example provided in the text is based on Bon [1986b]. The Appendix is based on Bon [1987]. Note that the question whether to undertake an economic action sooner or later is fundamental to economic behavior in general. (See note 5 in Chap. 1.)

18. As Hayes [1979: 265, emphasis in the original] put it, *"[w]e are never going to get an adequate formalization of common-sense by making short forays into small areas, no matter how many of them we make."*

19. In the United Kingdom, the life-cycle costing analysis has been popularized by Peter A. Stone [1980 and 1983], under the name of costs-in-use technique. In the United States, Harold E. Marshall and his associates from the National Bureau of Standards have accomplished much in this area. Under Marshall's guidance, the American Society for Testing and Materials has produced several standards regarding investment criteria applied to buildings and building systems. See, for example, "Standard Practice for Measuring Life-Cycle Costs of Buildings and Building Systems," ASTM, E 917–83, October 1983.

20. This section and the section that follows are based on the ongoing research on real property portfolio management at the Laboratory of Architecture and Planning, MIT. Since 1983 our sponsors were the Office of Facilities Management, Division of Capital Operations and Planning, Commonwealth of Massachusetts; Real Estate and Construction Division, International Business Machines Corporation; Construction Engineering Research Laboratory, U.S. Army Corps of Engineers; Institute of Technology, Shimizu Construction Company; and Corporate Property Management, Digital Equipment Corporation. For an overview of this research see Bon, Joroff, and Veale [1987], and *The MIT Report* [1987]. Also, Bon [1988] discusses some real property portfolio management issues in the context of building maintenance and replacement.

21. For surveys of real property management practices in the United States, including the structure of real property holdings of private and public organizations, see Zeckhauser and Silverman [1981 and 1983] and Veale [1988a and 1988b].

22. For a related discussion, concerning the time-profile of owned and leased space, see Levi and Matz [1987], for example.

23. For an introduction to industrial engineering and quality control techniques applied to construction, see Hashimoto [1986], for example.

24. The building client sometimes decides for new construction before carefully considering space utilization. As Hillebrandt [1984: 50] observes, "[the client] may want more usable space but this may be achievable by a rearrangement of existing spaces rather than by an extension of the premises." This is an important issue in connection with capital utilization and capital investment in general.

25. In this context, it is perhaps worth emphasizing that commonsense should not be confused with "common wisdom." An individual may, in fact, undertake a particular course of action in search for profit opportunities precisely because it runs counter to

the collective judgement. A spectacular example of such speculative action is afforded by Donald J. Trump, who "gambled" in mid-1970s that New York City would rebound from the debilitating effects of a national recession by initiating a series of projects that were perceived by the New York City real estate magnates as outright foolhardy. The subsequent events proved Trump to be a man of vision. As Tuccille [1985] shows, Trump's success depended in a crucial way on the erroneous collective judgement of real estate developers about the fate of New York City.

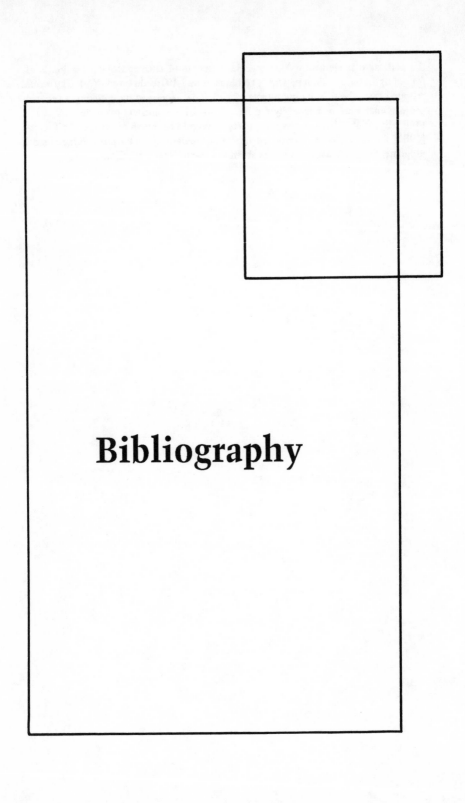

Bibliography

ARNOLD III, JASPER H. [1986], "Assessing Capital Risk: You Can't Be Too Conservative," *Harvard Business Review*, September-October, pp. 113–21.

BARRIE, DONALD S., ed. [1981], *Directions in Managing Construction: A Critical Look at Present and Future Industry Practices, Problems, and Policies*, New York: Wiley.

BARRIE, DONALD S., and BOYD C. PAULSON, JR. [1978], *Professional Construction Management*, New York: McGraw-Hill.

BAUMOL, WILLIAM J. [1970], *Economic Dynamics: An Introduction*, New York: Macmillan.

BAUMOL, WILLIAM J. [1977], *Economic Theory and Operations Analysis, Fourth Edition*, Englewood Cliffs, New Jersey: Prentice-Hall.

BERGSON, HENRI [1910], *Time and Free Will*, London: George Allen & Unwin.

BETANCOURT, ROGER R., and CHRISTOPHER K. CLAGUE [1981], *Capital Utilization: A Theoretical and Empirical Analysis*, Cambridge and New York: Cambridge University Press.

BIERMAN, Jr., HAROLD, and SEYMOUR SMIDT [1984], *The Capital Budgeting Decision: Economic Analysis of Investment Projects, Sixth Edition*, New York and London: Macmillan.

BISCHOFF, CHARLES W., and EDWARD C. KOKKELENBERG [1987], "Capacity Utilization and Depreciation-in-Use," *Applied Economics*, vol. 19, pp. 995–1007.

BOBROW, DANIEL G., ed. [1985], *Qualitative Reasoning about Physical Systems*, Cambridge, Massachusetts: The MIT Press (first published in 1984).

BÖHM-BAWERK, EUGEN [1959a], *History and Critique of Interest Theories*, Vol. 1, *Capital and Interest*, South Holland, Illinois: Libertarian Press (first published in 1884).

BÖHM-BAWERK, EUGEN [1959b], *Positive Theory of Capital*, Vol. 2, *Capital and Interest*, South Holland, Illinois: Libertarian Press (first published in 1889).

BÖHM-BAWERK, EUGEN [1959c], *Further Essays on Capital and Interest*, Vol 3, *Capital and Interest*, South Holland, Illinois: Libertarian Press (first published in two parts in 1909 and 1912).

BON, RANKO [1986a], "Choices, Values, and Time: The Psychology of Cost-Benefit Analysis," *Building Research & Practice*, vol. 14, no. 4, pp. 223–25.

BON, RANKO [1986b], "The Naive Building Economics Manifesto," *Building Research & Practice*, vol. 14, no. 6, pp. 348–51.

BON, RANKO [1986c], "Timing of Space: Some Thoughts on Building Economics," *Habitat International*, vol. 10, no. 4, pp. 93–99.

BON, RANKO [1987], "Expert Systems, Qualitative Reasoning, and Naive Building Economics," in P. Brandon, ed., *Computers and Building Cost Modelling*, London: Spon, pp. 419–28.

BON, RANKO [1988], "Replacement Simulation Model: A Framework for Building Portfolio Decisions," *Construction Management and Economics*, vol. 6, no. 2, Summer, pp. 149–59.

BON, RANKO, MICHAEL L. JOROFF, and PETER R. VEALE [1987], *Real Property Portfolio Management*, Discussion Paper and Symposium Summary, Real Property Portfolio Management Symposium, Laboratory of Architecture and Planning, MIT, Cambridge, Mass.

BROMILEY, PHILIP [1986], *Corporate Capital Investment: A Behavioral Approach*, Cambridge: Cambridge University Press.

BUCHANAN, JAMES M. [1969], *Cost and Choice: An Inquiry in Economic Theory*, Chicago and London: The University of Chicago Press (Midway Reprint, 1978).

CASSIMATIS, PETER J. [1969], *Economics of the Construction Industry*, Studies in Business Economics, No. 111, New York: National Industrial Conference Board.

COLLIER, COURTLAND A., and DON A. HALPERIN [1984], *Construction Funding: Where the Money Comes From, Second Edition*, New York: Wiley (first published in 1974).

DE KLEER, JOHAN, and JOHN S. BROWN [1985], "A Qualitative Physics Based on Confluences," in Daniel G. Bobrow, ed., *Qualitative Reasoning about Physical Systems*, Cambridge, Massachusetts: The MIT Press, pp. 7–83.

DENISON, EDWARD F. [1980], "The Contribution of Capital to Economic Growth," *American Economic Review*, vol. 70, pp. 220–24.

DOLAN, DENNIS F. [1979], *The British Construction Industry: An Introduction*, London and Basingstoke: Macmillan.

DOLAN, EDWIN G., ed. [1976], *The Foundations of Modern Austrian Economics*, Kansas City: Sheed & Ward.

DRUCKER, PETER F. [1985], *Managing in Turbulent Times*, New York: Harper Colophon Books (first published in 1980).

DYSON, FREEMAN J. [1988], *Infinite in All Directions*, New York: Harper & Row.

ECCLES, ROBERT G. [1981], "The Quasifirm in the Construction Industry," *Journal of Economic Behavior and Organization*, vol. 2, pp. 335–57.

EPLEY, DONALD R., and JAMES A. MILLAR [1984], *Basic Real Estate Finance and Investment, Second Edition*, New York: Wiley.

FELDSTEIN, MARTIN, ed. [1987], *Taxes and Capital Formation*, A National Bureau of Economic Research Project Report, Chicago and London: The University of Chicago Press.

FISHER, IRVING [1965], *The Theory of Interest*, New York: Sentry Press (first published in 1930).

FORBUS, KENNETH D. [1985], "Qualitative Process Theory," in Daniel G. Bobrow, ed., *Qualitative Reasoning about Physical Systems*, Cambridge, Mass.: The MIT Press, pp. 85–168.

FOSS, MURRAY F. [1963], "The Utilization of Capital Equipment," *Survey of Current Business*, vol. 43, pp. 8–16.

FOSS, MURRAY F. [1984], *Changing Utilization of Fixed Capital: An Element in Long-Term Growth*, Washington, D.C., and London: American Enterprise Institute for Public Policy Research.

GALBRAITH, JOHN K. [1984], "From Economics to Architecture and the Arts: A Journey," *Transactions 5*, The record of papers presented to the Royal Institute of British Architects, vol. 3, no. 1, pp. 57–63.

GARRISON, ROGER W. [1985], "A Subjectivist Theory of a Capital-using Economy," Chap. 8 in Gerald P. O'Driscoll, Jr., and Mario J. Rizzo, *The Economics of Time and Ignorance*, Oxford and New York: Basil Blackwell, pp. 160–87.

GEORGESCU-ROEGEN, NICHOLAS [1971], *The Entropy Law and the Economic Process,* Cambridge, Mass.: Harvard University Press.

GOLDTHWAITE, RICHARD A. [1980], *The Building of Renaissance Florence: An Economic and Social History,* Baltimore and London: Johns Hopkins University Press.

GRASSL, WOLFGANG, and BARRY SMITH, eds. [1986], *Austrian Economics: Historical and Philosophical Background,* New York: New York University Press.

GROÀK, STEVEN [1983], "Building Process and Technological Choice," *Habitat International,* vol. 7, nos. 5–6, pp. 357–66.

HABRAKEN, N. JOHN [1983], *Transformations of the Site,* Cambridge, Mass.: Awater Press.

HASHIMOTO, YOSHITSUGU [1986], *Improving Productivity in Construction Through QC and IE,* Tokyo: Asian Productivity Organization.

HAYEK, FRIEDRICH A. [1941], *The Pure Theory of Capital,* Chicago: The University of Chicago Press (Midway reprint, 1975).

HAYEK, FRIEDRICH A. [1948], *Individualism and Economic Order,* Chicago: The University of Chicago Press (Midway reprint, 1980).

HAYEK, FRIEDRICH A. [1975], *Profits, Interest and Investment,* Clifton, New Jersey: Augustus M. Kelley (first published in 1939).

HAYEK, FRIEDRICH A. [1984], *Money, Capital, and Fluctuations: Early Essays,* Chicago: The University of Chicago Press.

HAYES, PATRICK J. [1979], "The Naive Physics Manifesto," in Donald Michie, ed., *Expert Systems in the Microelectronic Age,* Edinburgh: Edinburgh University Press, pp. 242–70.

HAYES, ROBERT H., and DAVID A. GARVIN [1982], "Managing as if Tomorrow Mattered," *Harvard Business Review,* vol. 60, no. 3, pp. 71–79.

HICKS, JOHN R. [1946], *Value and Capital, Second Edition,* Oxford: Clarendon Press.

HICKS, JOHN R. [1965], *Capital and Growth,* Oxford: Oxford University Press.

HICKS, JOHN R. [1973], *Capital and Time: A Neo-Austrian Theory,* Oxford: Clarendon Press.

HICKS, JOHN R. [1977], *Economic Perspectives: Further Essays on Money and Growth,* Oxford: Clarendon Press.

HICKS, JOHN R. [1979], *Causality in Economics,* Oxford: Blackwell.

HICKS, JOHN R. [1984], *The Economics of John Hicks,* Selected by Dieter Helm, Oxford and New York: Basil Blackwell.

HICKS, JOHN R., and W. WEBER, eds. [1973], *Carl Menger and the Austrian School of Economics,* Oxford: Clarendon Press.

HILLEBRANDT, PATRICIA M. [1974], *Economic Theory and the Construction Industry,* London and Basingstoke: Macmillan.

HILLEBRANDT, PATRICIA M. [1984], *Analysis of the British Construction Industry,* London: Macmillan.

HINES, Jr, JAMES R. [1987], "The Tax Treatment of Structures," in Martin Feldstein, ed., *Taxes and Capital Formation,* A National Bureau of Economic Research Project Report, Chicago and London: The Chicago University Press, pp. 37–50.

KALDOR, NICHOLAS [1985], *Economics Without Equilibrium,* Armonk, New York: M.E. Sharpe.

KALECKI, MICHAL [1971], *Selected Essays on the Dynamics of the Capitalist Economy,* Cambridge: Cambridge University Press.

KESSEL, REUBEN A., and ARMEN A. ALCHIAN [1962], "Effects of Inflation," *Journal of Political Economy,* vol. 70, pp. 521–37 (reprinted in *Economic Forces at Work: Selected Works of Armen A. Alchian,* Indianapolis: Liberty Press, 1977).

KEYNES, JOHN M. [1964], *The General Theory of Employment, Interest, and Money,* New York: Harcourt, Brace & World (first published in 1936).

KIRZNER, ISRAEL M. [1966], *An Essay on Capital,* New York: Augustus M. Kelley.

KIRZNER, ISRAEL M. [1976a], *The Economic Point of View: An Essay in the History of Economic Thought, Second Edition,* Kansas City: Sheed and Ward (first published in 1960).

KIRZNER, ISRAEL M. [1976b], "The Theory of Capital," in Edwin G. Dolan, ed., *The Foundations of Modern Austrian Economics,* Kansas City: Sheed & Ward, pp. 133–44.

KIRZNER, ISRAEL M. [1976c], "Ludwig von Mises and the Theory of Capital and Interest," in Laurence S. Moss, ed., *The Economics of Ludwig von Mises: Toward a Critical Reappraisal,* Kansas City: Sheed & Ward, pp. 51–65.

KIRZNER, ISRAEL M. [1978], "The Entrepreneurial Role in Menger's System," *Atlantic Economic Journal,* vol. 6, no. 3, pp. 31–45.

KIRZNER, ISRAEL M. [1979], *Perception, Opportunity, and Profit: Studies in the Theory of Entrepreneurship,* Chicago and London: The University of Chicago Press.

KNIGHT, FRANK H. [1934], "The Nature of Economic Science in Some Recent Discussion," *American Economic Review,* vol. 24, no. 2, pp. 226–37.

KOWALSKI, ROBERT [1979], *Logic for Problem Solving,* New York: North-Holland.

KREGEL, J.A. [1976], *Theory of Capital,* London and Basingstoke: Macmillan.

KUZNETS, SIMON [1961], *Capital in the American Economy: Its Formation and Financing,* Princeton, New Jersey: Princeton University Press for the National Bureau of Economic Research.

LACHMANN, LUDWIG M. [1976a], "On Austrian Capital Theory," in Edwin G. Dolan, ed., *The Foundations of Modern Austrian Economics,* Kansas City: Sheed & Ward, pp. 145–51.

LACHMANN, LUDWIG M. [1976b], "Toward a Critique of Macroeconomics," in Edwin G. Dolan, ed., *The Foundations of Modern Austrian Economics,* Kansas City: Sheed & Ward, pp. 152–59.

LACHMANN, LUDWIG M. [1977], *Capital, Expectations, and the Market Process: Essays on the Theory of the Market Economy,* Kansas City: Sheed Andrews & McMeel.

LACHMANN, LUDWIG M. [1978a], *Capital and Its Structure, Second Edition,* Kansas City: Sheed, Andrews & McMeel (first published in 1956).

LACHMANN, LUDWIG M. [1978b], "An Austrian Stocktaking: Unsettled Questions and Tentative Answers," in Louis M. Spadaro, ed., *New Directions in Austrian Economics,* Kansas City: Sheed Andrews & McMeel, pp. 1–18.

LACHMANN, LUDWIG M. [1978c], "Carl Menger and the Incomplete Revolution of Subjectivism," *Atlantic Economic Journal,* vol. 6, no. 3, pp. 57–59.

LACHMANN, LUDWIG M. [1982], "Ludwig von Mises and the Extension of Subjectivism," in Israel M. Kirzner, ed., *Method, Process, and Austrian Economics: Essays in Honor of Ludwig von Mises,* Lexington, Mass., and Toronto: Lexington Books, pp. 31–40.

LACHMANN, LUDWIG M. [1986], *The Market as an Economic Process,* Oxford and New York: Basil Blackwell.

LANGE, JULIAN E., and DANIEL Q. MILLS, eds. [1979], *The Construction Industry: Balance Wheel of the Economy,* Lexington, Mass., and Toronto: Lexington Books.

LEVI, GERALD M., and ELLIOT S. MATZ [1987], "Choosing Strategic Objectives," Chap. 2 in Robert A. Silverman, ed., *Corporate Real Estate Handbook: Strategies for Improving Bottom-Line Performance,* New York: McGraw-Hill, pp. 17–32.

LEWIN, PETER [1986], "Economic Policy and the Capital Structure," in Israel M. Kirzner, ed., *Subjectivism, Intelligibility and Economic Understanding,* Essays in Honor of Ludwig M. Lachmann on his Eightieth Birthday, New York: New York University Press, pp. 211–20.

LYNCH, KEVIN [1972], *What Time This Place?* Cambridge, Mass.: The MIT Press.

MADDISON, ANGUS [1987], "Growth and Slowdown in Advanced Capitalist Economies," *Journal of Economic Literature,* vol. 25, no. 2, pp. 649–98.

MENGER, CARL [1981], *Principles of Economics,* New York and London: New York University Press (first published in 1871; first English translation published in 1950).

MENGER, CARL [1985], *Investigations Into the Method of the Social Sciences With Special Reference to Economics,* New York and London: New York University Press (first published in 1883; first English translation published in 1963, under the title *Problems of Economics and Sociology*).

MILL, JOHN S. [1973], *Principles of Political Economy, with Some of Their Applications to Political Philosophy,* New York: Augustus M. Kelley (first published in 1848).

MISES, LUDWIG [1966], *Human Action: A Treatise on Economics, Third Revised Edition,* Chicago: Contemporary Books (first published in 1949).

MISES, LUDWIG [1976], *Epistemological Problems of Economics,* New York and London: New York University Press (first published in 1933).

MISES, LUDWIG [1978], *The Ultimate Foundation of Economic Science: An Essay on Method,* Kansas City: Sheed Andrews & McMeel (first published in 1962).

MISES, LUDWIG [1980], *The Theory of Money and Credit,* Indianapolis: Liberty (first published in 1912; first English translation published in 1934).

The MIT Report [1987], "Real Property Portfolio Management," vol. 15, no. 7, pp. 9–10.

MITTRA, SITANSU S. [1986], *Decision Support Systems: Tools and Techniques,* New York: Wiley.

MOORE, GEOFFREY H. [1983], *Business Cycles, Inflation, and Forecasting, Second Edition,* National Bureau of Economic Research Studies in Business Cycles, No. 24, Cambridge, Mass.: Ballinger (first published in 1980).

MOSS, LAURENCE S., ed. [1976], *The Economics of Ludwig von Mises: Toward a Critical Reappraisal,* Kansas City: Sheed and Ward.

MYRDAL, GUNNAR [1939], *Monetary Equilibrium*, London: William Hodge (first published in 1931).

NEWTON-SMITH, W.H. [1980], *The Structure of Time*, London, Boston, and Henley: Routledge & Kegan Paul.

NISBET, JAMES [1961], *Estimating and Cost Control*, London: B.T. Batsford.

O'DRISCOLL, JR., GERALD P., and MARIO J. RIZZO [1985], *The Economics of Time and Ignorance*, Oxford and New York: Basil Blackwell.

PAULSON, Jr., BOYD C. [1981], "Research and Development," in Donald S. Barrie, ed., *Directions in Managing Construction: A Critical Look at Present and Future Industry Practices, Problems, and Policies*, New York: Wiley, pp. 407–33.

POWELL, CHRISTOPHER G. [1980], *An Economic History of the British Building Industry: 1815-1979*, London: The Architectural Press.

POWELL, CHRISTOPHER G. [1987], "Quest for Quality: Some Attributes of Buildings Affecting Judgement of Quality," *Design Studies*, vol. 8, no. 1, pp. 26–32.

RICARDO, DAVID [1951], *On the Principles of Political Economy and Taxation*, Vol. 1 in Piero Sraffa with Maurice H. Dobb, eds., *The Works and Correspondence of David Ricardo*, Cambridge: Cambridge University Press (first published in 1817).

ROSEFIELDE, STEVEN, and DANIEL Q. MILLS [1979], "Is Construction Technologically Stagnant?" in Julian E. Lange and Daniel Q. Mills, eds., *The Construction Industry: Balance Wheel of the Economy*, Lexington, Mass., and Toronto: Lexington Books, pp. 83–114.

SAMUELSON, PAUL A. [1967], *Economics: An Introductory Analysis, Seventh Edition*, New York: McGraw-Hill.

SCHUMPETER, JOSEPH A. [1954], *History of Economic Analysis*, New York: Oxford University Press.

SHACKLE, GEORGE L.S. [1967], *The Years of High Theory: Invention and Tradition in Economic Thought, 1926-1939*, Cambridge: Cambridge University Press.

SHACKLE, GEORGE L.S. [1969], *Decision Order and Time in Human Affairs*, Cambridge: Cambridge University Press.

SHACKLE, GEORGE L.S. [1972], *Epistemics and Economics: A Critique of Economic Doctrines*, Cambridge: Cambridge University Press.

SHUTT, R.C. [1982], *Economics for the Construction Industry*, London and New York: Longman.

SIMON, HERBERT A. [1981], *The Sciences of the Artificial, Second Edition*, Cambridge, Mass., and London: The MIT Press (first published in 1969).

SIMON, HERBERT A. [1982], *Models of Bounded Rationality: Behavioral Economics and Business Organization*, Vol. 2, Cambridge, Mass., and London: The MIT Press.

SOLOW, ROBERT M. [1963], *Capital Theory and the Rate of Return*, Amsterdam: North-Holland.

SPADARO, LOUIS M. [1978], *New Directions in Austrian Economics*, Kansas City: Sheed, Andrews and McMeel.

STONE, PETER A. [1980], *Building Design Evaluation: Costs-in-Use, Third Edition*, London and New York: Spon (first published in 1967).

STONE, PETER A. [1983], *Building Economy: Design, Production and Organization—A Synoptic View, Third Edition,* Oxford: Pergamon Press (first published in 1966).

TEMPELMANS PLAT, HERMAN [1982], "Micro-Economic Analysis of the Process of Design, Construction, and Operation of Houses," *IABSE Journal,* vol. 14, pp. 1–14.

TEMPELMANS PLAT, HERMAN [1984], "Costs and the Responsibilities of Tenants and Housing Authorities," Proceedings of the Third International Symposium on Building Economics, Vol. 5, Working Commission W.55 on Building Economics, International Council for Building Research Studies and Documentation (CIB), Ottawa, July 18-20, pp. 83–93.

TUCCILLE, JEROME [1985], *Trump: The Saga of America's Most Powerful Real Estate Baron,* New York: Donald I. Fine.

TURIN, DUCCIO A. [1966], *What Do We Mean By Building?,* An inaugural lecture delivered at University College London on 14 February 1966, London: H.K. Lewis.

VEALE, PETER R. [1988a], "Corporate Real Estate Asset Management in the United States," Master's Thesis, Department of Architecture, MIT, Cambridge, Mass..

VEALE, PETER R. [1988b], *Managing Corporate Real Estate Assets: A Survey of U.S. Real Estate Executives,* Laboratory of Architecture and Planning, MIT, Cambridge, Mass..

VENTRE, FRANCIS T. [1982], "Building in Eclipse, Architecture in Secession," *Progressive Architecture,* December, pp. 58–61.

WARD, JR., ROBERTSON [1987], "Office Building Systems Performance and Functional Use Costs," Proceedings of the Fourth International Symposium on Building Economics, Vol. A, Working Commission W.55 on Building Economics, International Council for Building Research Studies and Documentation (CIB), Copenhagen, September 14-17, 1987, pp. 113–24.

WATERMAN, DONALD A. [1986], *A Guide to Expert Systems,* Reading, Mass.: Addison-Wesley.

WHITEHEAD, ALFRED N. [1969], *An Introduction to Mathematics,* Oxford: Oxford University Press (first published in 1911).

WICKSELL, KNUT [1934], *Lectures on Political Economy,* Vol. I, London (first published in 1923).

WILLIAMSON, OLIVER E. [1975], *Markets and Hierarchies: Analysis of Antitrust Implications,* New York: Free Press.

WILLIAMSON, OLIVER E. [1979], "Transaction Cost Economics: The Governance of Contracting Relations," *Journal of Law and Economics,* vol. 22, pp. 233–61.

WINSTON, GORDON C. [1982], *The Timing of Economic Activities: Firms, Households and Markets, and Time-Specific Analysis,* Cambridge and New York: Cambridge University Press.

ZECKHAUSER, SALLY, and ROBERT A. SILVERMAN [1981], *Corporate Real Estate Asset Management in the United States,* A Report Prepared by Harvard Real Estate, Inc., in Cooperation With the National Association of Corporate Real Estate Executives, Cambridge, Mass..

ZECKHAUSER, SALLY, and ROBERT A. SILVERMAN [1983], "Rediscover Your Company's Real Estate," *Harvard Business Review,* January-February, pp. 111–17.

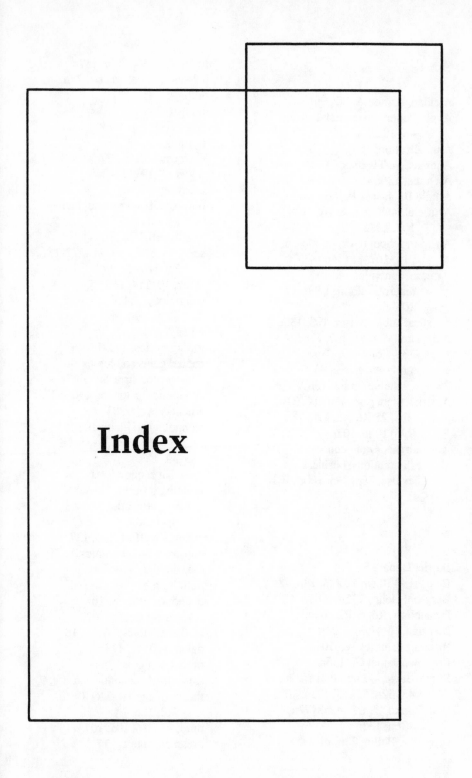

Index